African Cichlids of Lakes Malawi and Tanganyika

DEDICATION

This book is dedicated to Peter and Henny Davies of Cape McClear, Lake Malawi for all the beautiful fishes they found and made available for aquarists.

African Cichlids of Lakes Malawi and Tanganyika

By Dr. Herbert R. Axelrod and Dr. Warren E. Burgess

Pseudotropheus elongatus (or at least of the *elongatus* complex). Photo by Andre Roth.

Eleventh Edition

This edition contains an additional 32 pages of full-color photos (covering both Lake Malawi and Lake Tanganyika) that did not appear in the tenth edition. These photos appear beginning on page 353.

Table of Contents

Photo survey of Lake Malawi Cichlids starts
on p. 122
Photo survey of Lake Tanganyika cichlids
starts on p. 296

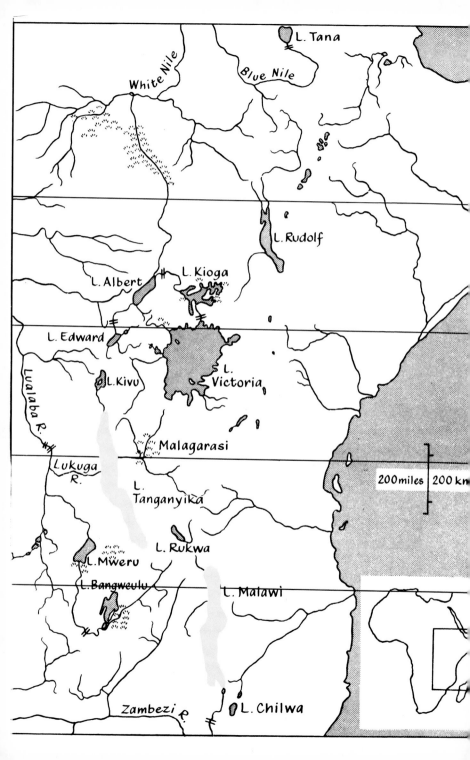

L. Tana

White Nile

Blue Nile

L. Rudolf

L. Albert   L. Kioga

L. Edward

Lualaba R.

L. Kivu   L. Victoria

Malagarasi

Lukuga R.

L. Tanganyika

L. Rukwa

L. Mweru

L. Bangweulu   L. Malawi

Zambezi R.   L. Chilwa

200 miles | 200 km

Introduction

Back Row: Dr. Warren E. Burgess, Dr. Herbert R. Axelrod, Glen S. Axelrod. Front Row: Henny and Peter Davies. Photo by Evelyn Axelrod.

The great lakes of East Africa, with the major associated river systems, and barriers—waterfalls and rapids—which have been and are important in the isolation of various basins. The map and caption were taken from Fryer and Iles' *The Cichlid Fishes of the Great Lakes of Africa.*

There have been many expeditions to the great lakes of Africa since their discovery. Some of the first cichlid specimens were sent back to Europe by the famous Dr. David Livingstone, contributor to the discovery and explorations of the lakes, and Dr. John Kirk (a botanist with the second Livingstone expedition). Subsequent expeditions and collections were made by people such as Prof. J. E. S. Moore, Dr. W. A. Cunnington, Dr. L. Stappers, Mr. R. Wood, and Dr. C. Christy. These names and others are familiar to African lake cichlid enthusiasts in the names of their favorite fishes (livingstoni, kirki, moorei, cunningtoni, stappersi, woodi, christyi, etc.). The cichlids they collected and sent back for study were worked on by such famous ichthyologists as Dr. Albert Guenther, Dr. G. A. Boulenger, Dr. C. Tate Regan, and more recently by Dr. E. Trewavas, Dr. P. H. Greenwood, and Dr. Max Poll. A definitive book on the biology and evolution of these cichlids (The Cichlid Fishes of the Great Lakes of Africa) was written by Drs. Fryer and Iles.

In recent years these lake cichlids exploded on the aquarium scene, causing quite a stir. Scientists were hard put to keep up with the new importations, identifying some as previously described species and determining that others were new to science. The demand remained high in spite of the high initial cost of these cichlids and has continued now that local breeders and hobbyists are turning out enough lake cichlids to satisfy much of the demand and bring the prices down.

The major sources of these cichlids are Lakes Malawi and Tanganyika. It was mainly due to the efforts of the early collectors and shippers that we were able to obtain the number and variety of species we have today. They had to overcome a great number of obstacles, often working under very primitive conditions as well as working with fishes about which virtually nothing was known. In Lake Tanganyika Pierre Brichard and his family were the pioneers, and it is a rare fish from that lake that does not owe its presence in European and American markets to the Brichards' efforts. Even domestically raised fishes probably had their "ancestors" shipped from the lake by Pierre Brichard. In

Lake Malawi has islands and peninsulas (such as this one) which provide extra shoreline and therefore habitats for the many cichlids found there. Photo by Dr. Herbert R. Axelrod.

Lake Malawi it was Peter Davies and his wife Henny although two other major shippers, Mr. Stewart Grant and Eric Fleet, contributed much to supply Lake Malawi cichlids to Europe and America. Since the Davieses' departure to England it is the latter gentlemen who have continued to provide the new and interesting cichlids from the lake.

Many of the species of lake cichlids have become firmly established in the aquarium hobby, and local breeders are now able to supply almost enough to meet the demand. The prices have dropped from astronomical to reasonable, so the average hobbyist can afford a pretty good variety of African lake cichlids. The knowledge of these fishes has also grown, to the extent that breeding them is relatively easy and losses are minimal. Does that mean that we should stop looking to the importers for the lake cichlids? Certainly not! Between the two lakes there are close to 400 species of cichlids, the vast majority of which are suitable for keeping in home aquaria. Add to this the great variety of morphs of the various fishes and you can see that there is much more to come. It is our hope to be able to keep the reader as much up to date as possible with the knowledge and identification of these cichlids from Lakes Malawi and Tanganyika.

The Lakes | 1

Aerial view of Cape Maclear, Lake Malawi, about 16 kilometers from Monkey Bay. Photo by Dr. Herbert R. Axelrod.

Tilapia rendalli is one of the few species occurring in both Lake Malawi and Lake Tanganyika. Photo by J. Voss, Universite de Liege.

15

Near the end of the Pliocene a tremendous increase in volcanic activity and shifting of the earth's surface resulted in the formation of the African rift valleys. It was within these rifts that the great lakes were formed. In what is called the Western Rift the northernmost of these lakes is Lake Albert, followed, as one progresses southward, by Lakes Edward and George, Lake Kivu, and finally Lake Tanganyika. Southeast of Lake Tanganyika, also in the Western Rift, is Lake Rukwa. There are a number of smaller lakes in the northern part of the Eastern Rift, the most prominent of which is Lake Rudolf. Further south in this same rift can be found Lake Malawi (formerly Lake Nyasa). Between these two major rifts is the shallow but very large Lake Victoria, and to the southwest of the rifts are two additional lakes, Mweru and Bangweulu. Other less important lakes are scattered through the area.

The three largest lakes, Lake Malawi, Lake Tanganyika, and Lake Victoria, are among the top ten lakes in the world as far as size is concerned. Lake Victoria, with an area of about 68,800 square kilometers, ranks third in the world, behind only the Caspian Sea and Lake Superior. It is about 320 kilometers long and almost as broad but has a depth of only 83-93 meters at its deepest points. Lake Tanganyika, the seventh largest lake in the world, has an area of about 34,000 km², a length of some 675 km and a depth of 1,470 meters, making it the second deepest lake (Lake Baikal is the deepest). Lake Malawi is of considerable size also, being the ninth largest lake and having an area of 30,000 km², a length of 580 kilometers, and a depth of 704 meters. The two rift lakes (Malawi and Tanganyika) are long and narrow but very deep; Lake Victoria, not a rift lake, has a much larger surface area but is relatively shallow.

Among the more interesting features of the lakes—and one that is very important as far as aquarists are concerned—is the alkalinity of their water. Aquarists are used to pH values around the neutral point or slightly acid for most of the species of fishes kept. But these lakes have pH values well above neutral; i.e., Malawi 7.7-8.6, Tanganyika 8.6-9.2, and Victoria 7.1-9.0. This

16

Not all parts of the Lake Malawi shores are rocky. Certain areas, shown in the photos above and below, are sandy or with vegetation. Most of the more colorful fishes are found, however, in the rocky areas, where they are much more difficult to capture. Photos by Dr. Herbert R. Axelrod.

Most species keep rather close to the rocks for feeding and protection. Seen here are a group of Lake Malawi mbuna including *Melanochromis*, *Pseudotropheus*, and *Labeotropheus* species. Photo by Dr. Warren E. Burgess.

An underwater scene around the rocks in about 6 meters of depth. A dominant *Pseudotropheus zebra* BB male stands out among the other mbuna (mostly *Pseudotropheus* and *Melanochromis* species). Photo by Dr. Herbert R. Axelrod.

A *Pseudotropheus* sp. (probably *P. tropheops*) along with a pair of *Melanochromis* sp. over some large rocks in Lake Malawi. Photo by Dr. Warren E. Burgess.

A varied assortment of *Pseudotropheus* species (including several male BB *P. zebra*). When there is no threat of danger some fishes hover above the rocky surfaces, feeding on the drifting zooplankton of the upper layers. Photo by Dr. Herbert R. Axelrod.

The huge rocks in shallow water support a thick layer of biocover ("aufwuchs") which the cichlids feed upon. This photo was taken at a depth of 3-4 meters near Cape MacClear showing the cichlids busily feeding. Photo by Dr. Herbert R. Axelrod.

means that the water has to be especially prepared and that the lake cichlids must be kept by themselves, since few other fishes can survive in the very alkaline water. Another important feature of Lakes Malawi and Tanganyika is that they are in a state of more or less permanent stratification: the lower layers have been depleted of oxygen and therefore are unsuitable for the fishes. The deepest a cichlid has been found in Lake Tanganyika is only about 215 meters. For a lake that is over 1,400 meters deep, this depth represents only the upper surface waters. Most cichlids do not penetrate even that far and are more likely to be found in the warmer sunlit waters of 50 meters or less.

The rift lakes are generally isolated from one another and as such have allowed the process of evolution to proceed in its own direction in each lake. In non-cichlid groups the process seems to have proceeded slowly; in cichlid fishes there has been a tremendous explosive radiation, producing a great variety of species in a relatively short geologic time. In Lake Malawi, for example, there are 225 described species of cichlids and only forty-odd

Swamplands and marshy areas can also be found along the shores of Lake Malawi. The upper photo was taken at an altitude of about 600 meters near Monkey Bay; the lower photo is a closeup of this marshland. Photos by Dr. Herbert R. Axelrod.

A scene in a grassy habitat of Lake Malawi, taken in about 6 meters depth. A school of *Haplochromis* species (with vertical barring) can be seen in the background. Photo by Dr. Herbert R. Axelrod.

The edge of the grassy area in Lake Malawi where several predators can be seen, such as *Haplochromis johnstoni* (vertical bars) and *H. livingstonii* (blotchy fish). Photo by Dr. Herbert R. Axelrod.

Pseudotropheus tropheops in a rather exposed position for an mbuna. Perhaps its pale color makes it less visible to a predator. Photo by Dr. Herbert R. Axelrod.

A dominant male *Pseudotropheus zebra* and *Haplochromis* sp. (probably *H. epichorialis*) venturing out over the sand—but not too far from the rocks. Photo by Dr. Herbert R. Axelrod.

Pierre Brichard with an early model of his "fish compression chamber." Certain species (ex. *Cyphotilapia frontosa*) must be decompressed or else they would suffer from the "bends" and probably die. Photo by Dr. Herbert R. Axelrod.

Above: An aquarist's dream in only 5 meters of water in Lake Tanganyika. Easily recognizable are the school of *Lamprologus brichardi* and *Julidochromis marlieri.* Photo by Dr. Herbert R. Axelrod.

Below: Pierre Brichard searching among the rocks for Lake Tanganyikan fishes in about 10 meters of water. He uses an anesthetic (quinaldine dissolved in acetone or alcohol) to make the fishes groggy so they will be easier to catch. Photo by Dr. Herbert R. Axelrod.

Looking across Lake Tanganyika at the Zaire mountains from the Bujumbura, Burundi shore. Photo by Dr. Herbert R. Axelrod.

Lake Tanganyika is large enough to have a surf of sorts. Even in this habitat there are fishes, such as the goby cichlids. Photo by Glen S. Axelrod.

The cliffs of Mgu wa Tembo (which in Swahili means elephant's foot) mark the entrance to Kigoma Bay. Photo by Glen S. Axelrod.

A section of the shoreline of Lake Tanganyika taken from a plane about to land at Kigoma, Tanzania. Photo by Glen S. Axelrod.

The lake level fluctuates with the seasons, becoming lower during the dry season and rising during the rainy season. Photo by Dr. Herbert R. Axelrod.

species of non-cichlid fishes. In Lake Tanganyika the picture is the same, with 158 known species of cichlid fishes and another 67 species of non-cichlids. Lake Victoria follows the pattern, with about 175 cichlid species and 38 non-cichlids. What is so unusual about these species is that they are almost all endemic to their respective lakes. (An endemic species is one that is found in a particular (usually limited) geographic area and no other place in the world.)

Diving is relatively safe in these lakes, although the hippos and crocodiles may present a problem or two at times. One of the authors (HRA) had a close call with a huge electric catfish in Tanganyika, and his diving partner, Thierry Brichard, was almost bitten by a water cobra *(Boulengerina annulata)* which was coiled and ready to strike . . . underwater!

Not as visible as the hippos and crocodiles are the disease-producing digenetic trematodes. These parasitic organisms, usually of the genus *Schistosoma* (causing the disease schistosomiasis), go through a complicated life cycle which includes a free-swimming miracidium stage, two generations in a molluscan (usually a snail) host, and finally the cercarian stage that infects humans. Both Lakes Malawi and Tanganyika are relatively free from this parasite, although there are areas in both lakes that should be avoided; Lake Victoria, however, is heavily infected and can be a danger to divers.

The cichlids in these lakes are quite varied. In Lake Tanganyika, for example, the smallest cichlid is reported to be *Lamprologus multifasciatus*, which measures only about 35 mm; the largest is *Boulengerochromis microlepis,* which attains a length of over 900 mm. By weight the largest cichlid is more than 800 times heavier than the smallest.

Collecting the fishes in the lakes may be just as easy as setting a trap and checking it every once in a while to see what has been captured or as difficult as diving with SCUBA gear and hand nets and chasing very agile fishes around a group of rocks in deep water. In this latter case the fish anesthetic quinaldine, used to make the fish groggy while it is netted, helps increase the catch as well as causing less injury to the fish while netting. These relatively deep-water cichlids must also be brought up to the surface carefully, for they are subject to pressure problems in which the swim bladder expands (to the extent that it even protrudes from the mouth) and causes damage to the fish. Pierre Brichard devised a decompression chamber which alleviates this problem.

The extensive shorelines of both Lakes Malawi and Tanganyika provide suitable habitats for the vast array of different cichlid species. There are rocky areas, sandy areas, and grassy areas; each has its complement of fishes, although the rocky areas shelter the most species by far. It is also these rock-dwelling fishes that provide the basis for the vast majority of our aquarium fishes. In Malawi the natives have given these rock-dwelling cichlids a special name which has caught on with hobbyists—mbuna.

The supply of cichlids from these lakes is by no means exhausted. With further explorations many new species will be discovered and hopefully find their way into hobbyists' tanks. With the political situations rapidly changing in the area of the lakes, however, it is not known whether certain supplies will be cut off. There are enough stable governments in the area that it would seem that we will have a supply of Malawi and Tanganyika cichlids and—with the departure of Idi Amin—also Lake Victoria cichlids.

Rocks and Islands | 2

A deeper part of Lake Tanganyika, the typical habitat of *Cyphotilapia frontosa* (seen here). Photo by Dr. Herbert R. Axelrod.

Telmatochromis caninus is a rock-dwelling species from Lake Tanganyika. A cave or other shelter is usually selected as a spawning site. Photo by H.-J. Richter.

The great lakes of Africa have been likened to a group of islands. But where islands are bits of land surrounded by water, the lakes are bits of water surrounded by land. For animals on the islands or fishes in the lakes this means the same thing—isolation. The fishes of the various lakes are therefore isolated from each other, and the evolution within each lake proceeds independently in its own direction. The speed of the evolution of the species in a lake depends upon the plasticity of the species involved, the biological and physical pressures exerted on them, and the length of time these pressures have been acting upon the animals. The present evidence indicates that Lake Victoria is the youngest of the three largest lakes, Lake Malawi second oldest and Lake Tanganyika the oldest. The number of endemic genera appears to be correlated with the age of the lakes, with Lake Tanganyika (the oldest) having 42 genera (35 of which are endemic to the lake), Lake Malawi 27 genera (22 of which are endemic) and Lake Victoria (the youngest) with 8 genera (4 of which are endemic), whether by some causal factors or coincidence or both.

Even within a lake the "island" type evolutionary system seems to be at work. There are rocky areas scattered around the lakes separated from each other by more or less open sandy or grassy areas. The fishes living in a rocky area are effectively isolated from the fishes living in adjacent rocky areas by the simple fact that if they dared venture too far from their rocky shelter they would be at the mercy of the predators that live in these open areas. It is almost as if each fish had a rubber band stretched between itself and the rocks; as the fish moves farther away from the shelter of the rocks the "pull" gets stronger and stronger. This means that the breeding populations are restricted to their own particular rocky area and are free to go off in their own evolutionary direction independent of what is going on in other parts of the lake. Of course similar biological and physical pressures are acting on the fishes, and the general trends tend to be the same. But differences do occur and these differences, particularly variations in the color and pattern, can be seen in fishes

A dominant male *Pseudotropheus zebra* (BB) stands out prominently against the rock surfaces. This is probably a territorial signal warning other *P. zebra* to keep away or serving as an attractant to passing female *P. zebra*. Photo by Dr. Herbert R. Axelrod.

of the same species from different parts of the lake. Chances are that if you took the progeny from a single pair of cichlids and divided them between two suitable habitats but made sure they had no contact whatever, eventually they would start to show significant differences between the populations.

The isolation effect in the lakes is slowed by several factors, one of which is that the fluctuations of the level of the lake create new barriers where there once were none and remove some of the barriers that were there. Another factor is that there are some successful crossings of the barriers by a fish or group of fish.

The lakes may be divided into a number of different ecological zones, each supporting its own particular complement of species. These zones are the estuarine and swampy areas, the rocky areas, the open sandy or grassy patches, the deep benthic areas, and the pelagic or open water areas. The swampy or estuarine

Variations on a theme. These *Pseudotropheus zebra* males have the BB pattern but the fish in the upper photo looks very pale as the black bars have faded to a blue-gray. The individual in the lower photo is more normally colored. These fishes can control the pattern and intensity of color to some degree depending upon whether they are frightened, threatening, trying to attract a female, etc. Photos by Dr. Herbert R. Axelrod.

A sandy area in Lake Tanganyika showing two male *Cyathopharynx furcifer* fighting. The barred fish is *Cyphotilapia frontosa*. Photo by Dr. Herbert R. Axelrod.

Lamprologus tretocephalus standing guard over a brood of young in Lake Tanganyika. The young can be seen under the rock ledge. Photo by Glen S. Axelrod.

zone usually accommodates species that are adapted to the chemical and physical conditions of the river and will not (or cannot) enter the waters of the lake proper. The bottom material may be mud or sand or a mixture of the two, or in some areas there may be plants growing. *Haplochromis callipterus* is an example of a fish from such an area. As might be expected, the rocky zone supports the most species, as the shelter afforded by the rocks allows the inhabitants to live and breed in more secure conditions than in other zones. It is around the rocks that divers will see the most colorful species. *Pseudotropheus zebra* is an example. The open sandy stretches provide little or no protection and are inhabited by some of the large predators as well as species which are cryptically colored so that they can blend in with the background. *Haplochromis rostratus* is an example. It is interesting to note here that one of the species of *Pseudotropheus* (normally rock-dwelling species), *P. lanisticola*, lives in sandy areas but has become adapted to using abandoned snail shells for protection. Again, the sandy stretches may be interspersed with grassy areas which afford some protection. Predators will often patrol these open areas or lie camouflaged (usually by color and pattern) in wait for unsuspecting or wandering prey. The deep benthic zone is an area in which the number of fishes starts to thin out as the amount of oxygen begins to decrease. *Haplochromis heterotaenia* and species of the genus *Rhamphochromis* are examples. Finally there is the open water or pelagic zone, which supports the plankton feeders and the fishes that prey upon them. The plankton feeders (generally referred to as the utaka) are commonly silvery in color with some countershading, making them all but invisible in the open waters. They also tend to be schooling fishes as well, making up for their lack of protection by sheer numbers. *Haplochromis chrysonotus* is an example. Of course these zones often merge into one another but in general are fairly discrete.

The major problem of solving the mystery of the relationships and evolution of the lake cichlids is to obtain an adequate sample of the species as well as a thorough study of the ecology of the lakes themselves. The early expeditions obtained a large number of specimens which now reside in various museum collections around the world. These have been and are still being worked upon by scientists, as evidenced by the papers being published. Now we also have people at the lakes collecting live specimens for the aquarium hobby. As these new specimens arrive on the scene,

Most Lake Malawi mbuna keep relatively close to the rock surface and are ready to dart into the shelter of crevices or caves when danger threatens. Photo by Dr. Herbert R. Axelrod.

names by which they can be sold are demanded. But who can equate the brightly colored living individuals with the printed description in a journal or even actual preserved specimens! Eventually dead specimens are obtained by competent ichthyologists, and names become available for the aquarium species. Meanwhile, one or more trade names have been applied to the fish, thoroughly complicating matters as these fishes become distributed along with their false names. Some of the aquarium species even turn out to be new to science and must be described. Considering the confused status of the already described species, this is often not a simple matter. It is hoped that eventually these lake fishes will all be straightened out taxonomically; when they are, both hobbyists and scientists can give a long sigh of relief.

Peculiarities of African Lake Cichlids | 3

Haplochromis compressiceps is a predator on other fishes. Its very compressed body is difficult to see as it suspends itself head-down with its slender body facing the prey. Photo by Dr. Herbert R. Axelrod at the Berlin Aquarium.

The fleshy lips of *Haplochromis euchilus* are very unusual. They are suspected of being a feeding adaptation. Photo by Dr. Herbert R. Axelrod.

The explosive evolution of the African lake cichlids has produced a number of very interesting and unusual forms. In adapting to the various niches that were open to them when their ancestors first colonized the lakes, some of the cichlids have developed certain peculiarities. This includes species in both lakes which have developed the same form or structure even though they belong to different genera. This short discussion by no means exhausts the unusual features of the lake cichlids.

Lake Tanganyika houses not only the largest but also the smallest cichlid species in the two lakes. The largest, *Boulengerochromis microlepis*, appears also to be the largest cichlid in the world. It weighs in at more than 4.5 kilograms and has a length of about 900 mm when fully grown. The smallest cichlid is *Lamprologus multifasciatus*. It is reported to be mature at 25 mm and has a maximum length of no more than 35 mm. It lives in abandoned snail shells and apparently feeds on tiny invertebrates.

This habit of shell-dwelling may have developed because of a paucity of shelters. The rocky areas may have become too crowded or even too dangerous for the smaller species, and the open areas had little to offer other than the empty snail shells. Several species in both lakes have taken advantage of this unique "home" on a more or less permanent basis. In Lake Tanganyika *Lamprologus multifasciatus, L. ornatipinnis, L. wauthioni, L. ocellatus*, and perhaps *L. brevis* are shell-dwellers, as is *Pseudotropheus lanisticola* of Lake Malawi. Many more species from both lakes can be considered as temporary shell-dwellers, especially when considering the younger stages.

In Lake Tanganyika there are some species that act and to some extent look like the marine gobies and have therefore collectively been called goby cichlids. They are members of the genera *Eretmodus, Tanganicodus*, and *Spathodus*. These small stocky cichlids inhabit turbulent shoreline areas where they can be seen "hopping" among the smaller rocks, pebbles, and rubble.

In both lakes there are cichlids that are extremely laterally compressed. Lake Tanganyika, for example, has its *Lam-*

Compressed fishes are found in both Lake Malawi and Lake Tanganyika. Here is *Lamprologus compressiceps* from Lake Tanganyika (note the same descriptive specific name as the *Haplochromis* species from Lake Malawi). Photo by Wilhelm Hoppe.

prologus compressiceps, the specific name referring of course to the compressed condition of the fish. In Lake Malawi we have *Haplochromis compressiceps*. This latter species is a piscivore that lurks among the lake grasses, using the very thin body to avoid detection by the potential prey fishes. One can readily see the effect when looking head-on at one of these fish.

The development of thick expanded lips has occurred in several species of cichlids from Lake Malawi's *Haplochromis euchilus*, *H. ornatus*, and *H. festivus*, Lake Tanganyika's *Lobochilotes labiatus*, and Lake Victoria's *Haplochromis chilotes* to Central America's *Cichlasoma labiatum*. It is suspected that these expanded lips are sensory in nature and aid in the detection of food organisms. In the African lakes these fishes inhabit rocky areas where they feed by pressing their lips against a rock; once the food item is found, they pick it off with their simple teeth.

The genus *Labeotropheus* of Lake Malawi is characterized in part by the development of a projecting snout. It is not known what function this snout has—perhaps it is only to bring the mouth into a ventral position where it can be used for more efficient

Tropheus moorii from Lake Tanganyika is one of those species that have an unusually high number of color morphs. This is the Chaitka Cape variety or rainbow *Tropheus*. Photo by Pierre Brichard.

Simochromis dardennei is a silvery fish with dark markings (as shown) until it reaches a length of about 18 cm. It then goes into deeper water and takes on a completely different (and much more colorful) color pattern. Photo by Glen S. Axelrod.

Haplochromis tetrastigma lives in the more sandy areas of Lake Malawi. The pattern of stripes, bars, or, in this case, spots aid in the identification of the myriad of *Haplochromis* species in the lake. Photo by Dr. Herbert R. Axelrod.

Labeotropheus species (here *L. fuelleborni* is in the center) have a characteristic overhanging snout and straight mouth with all tricuspid teeth. Photo by Dr. Warren E. Burgess.

[A]

[B]

A side view of the Malawian "eye-biter," *Haplochromis compressiceps* (a) and a close-up of the head showing the compressed condition of the fish (b). Photos from Fryer and Isles' *The Cichlid Fishes of the Great Lakes of Africa.*

algal scraping. In Lake Tanganyika another fish, *Ophthalmochromis nasutus*, has a projecting snout. It is not as well developed as in the *Labeotropheus* species, and the only other species in the genus, *Ophthalmochromis ventralis*, lacks it. In fact males display a larger proboscis than the females and it seems to grow larger with age.

Two genera of cichlids in Lake Malawi and two in Lake Tanganyika are distinguished by having enlarged head pores. The names are very similar and have caused some confusion among hobbyists. In Lake Malawi the genera are *Trematocranus* and *Aulonocara*, whereas in Lake Tanganyika they are *Trematocara* and *Aulonocranus*. Some species of the genus

Haplochromis in Lake Malawi have enlarged head pores, as does *H. boops* from Lake Victoria, but these pores are not nearly as well developed as those of the first four genera. The function of these enlarged pores is not completely known. It has been theorized that they are an adaptation to a deep-water existence. *Trematocara* occurs in deep water (up to 200 meters according to Brichard), but *Aulonocranus dewindti* (the only species in the genus) is a shallow-water species, being found from the surface to a maximum of 6 meters, also according to Brichard. If these were nocturnal fishes the explanation could be that it is an adaptation to wandering about at night, but not enough is known about these fishes to support this theory.

A cephalic hump is not uncommon in cichlids, but there are two species, one from Lake Malawi and the other from Lake Tanganyika, that have exceptionally large humps. In Lake Malawi there is *Haplochromis moori*, which is basically blue in color although some dark markings show up from time to time depending upon the mood of the fish. *Cyphotilapia frontosa* is its counterpart from Lake Tanganyika; *C. frontosa* is white with several dark vertical bars.

What has attracted a great deal of attention in both lakes is the polymorphism of some of the species. In Lake Malawi the species with the most morphs seem to be concentrated in the mbuna. *Pseudotropheus zebra* might be called king of the morphs, with *P. tropheops* and the *Labeotropheus* species not too far behind. *Pseudotropheus zebra* comes in several different patterns such as dark vertical bars on a light background (called the BB morph) or orange with black blotches (OB) and in several solid colors (white, blue, red, orange, etc.). Variations on the basic pattern also occur. For example a BB with a reddish orange dorsal fin is called a red-top zebra. The barred individuals are generally males, the blotched ones females. There are males that are blotched, but they are quite rare and have a special name attached to them—marmalade cats.

In Lake Tanganyika the morph king is *Tropheus moorii*. This species exhibits variations in color and pattern including red, black, bronze, orange, yellow belly, rainbow, red-black, etc., as well as white band, olive band, narrow olive band, etc.

It is highly probable that not all of the morphs of either lake have been collected yet. This helps keep interest in these fishes high, as the next shipment might contain some new and brightly colored morphs.

How Cichlids Feed

A couple of mbuna (possibly *Petrotilapia tridentiger* and *Labeotropheus fuelleborni*) feeding on the rocky biocover. Notice the different angles of attack. Photo by Dr. Herbert R. Axelrod.

As seen here in this *Labeotropheus trewavasae,* the snout protrudes out over the jaws. This is thought to be a specialized feeding adaptation. Photo by Andre Roth.

The tremendous success of cichlids in the African lakes has been attributed to a large extent to their ability to take advantage of all the different food sources available from microscopic algae to fishes. The early colonizers of the lakes are thought to be riverine species that were more or less generalized feeders. From these ancestors evolved the various species you see now with specialized structures such as protrusible jaws, modified teeth for different types of prey organisms, streamlined or otherwise modified shapes, colors and/or patterns that help camouflage, etc. Most, if not all, of these structures have something to do with the efficiency with which these fishes are able to feed. A short discussion of the various feeding types from herbivore to carnivore follows.

Herbivorous fishes are those that feed on different types of plant material. This could be anything from microscopic filamentous algae or diatoms to the leaves of large plants. Among the cichlids that are known to utilize the microscopic floating plants (called phytoplankton) are species of *Trematocara, Ophthalmochromis, Ophthalmotilapia,* and *Sarotherodon/Tilapia.* There do not seem to be major modifications in the feeding structure of these fishes to efficiently tap this particular food source. In fact, the actual mechanisms of collecting and concentrating the food are insufficiently known, although some modifications of the sieving structure of the gill arches and production of mucus which may entrap the tiny plants may have occurred. Certain algae, for example the blue-greens, have cellulose walls that cannot be digested by the fishes. These algae pass through the entire gut of the fish and are excreted virtually undamaged. The diatoms are encased in a siliceous shell, but the shells are perforated with holes, enabling the fishes' digestive juices to get at the internal parts of the diatoms. They may also be crushed by the jaw or pharyngeal teeth.

Bottom deposits composed of sedimented planktonic algae are utilized by such fishes as species of the genus *Ectodus* as well as some of the fishes that feed on the floating algae. In some species of *Sarotherodon/Tilapia* the teeth are generally small but numerous, in several rows, and tricuspid.

The best known of the aquarium lake cichlids are probably what can be called algae scrapers or, in more scientific terms, epilithic algal feeders. The material generally grazed from the rock surface has been termed *aufwuchs;* this term is often broadened to include the entire biocover of the rocks, animal as well as plant. Here it will be restricted to the algal component. The jaws and teeth of the aufwuchs feeders have generally been modified in some way to facilitate this type of feeding. The teeth are usually numerous, in several rows, and bicuspid or tricuspid. The jaws are often straight so that a long row of teeth can be drawn across the surface of the rock at the same time. The various species of *Pseudotropheus,* for example *P. zebra* and *P. tropheops,* are good examples of this type feeder from Lake Malawi, *Tropheus* and *Simochromis* species are examples from Lake Tanganyika, and *Haplochromis nigricans* is an example from Lake Victoria. The outer row of teeth in these fishes is normally bicuspid; the rows backing them up are composed of tricuspid teeth. In some genera the bicuspid teeth are dispensed with and all of the scraping jaw teeth are tricuspid. This can be seen in Lake Malawi's *Petrotilapia tridentiger* and Lake Tanganyika's species of the genus *Petrochromis.* The teeth are on slender shafts which make them more flexible and adjustable to the irregular rock surfaces. The species of *Labeotropheus* also have tricuspid teeth in the jaws which are lined up in a straight line providing a continuous scraping surface. In addition the mouth is situated more ventrally, below a protruding snout. In the scapers the open mouth is placed against the rock; as the mouth closes the teeth are drawn across the surface, scraping the algae off as they go. This can be modified to a rapid nibbling process. In some species, for example *Haplochromis guentheri* of Lake Malawi, the scraping is abandoned in favor of selective picking of algal filaments on a more individual basis.

Since algae grow on the larger plants in the lake, the algal feeders can be seen working them over. *Cyathochromis obliquidens* of Lake Malawi and *Haplochromis lividus* of Lake Victoria seem equally proficient at removing algae from rock surfaces as well as the surface of plant leaves. *Hemitilapia oxyrhynchus,* on the other hand, seems more adapted to nibbling the algae growing on the plant leaves. It will maneuver to take a leaf in its jaws and nibble along its length, removing the algae but leaving the leaf itself undamaged in the process.

There are fishes that feed on the larger plants themselves. For

When a predator is around, such as this *Lamprologus elongatus* of Lake Tanganyika, it has the immediate vicinity all to itself. Most other fishes have taken shelter. Photo by Glen S. Axelrod.

Haplochromis tetrastigma of Lake Malawi is a bottom-feeder in the sandy areas. In this photo its mouth is positioned for feeding. Photo by Dr. Herbert R. Axelrod.

The scenes shown above and below are at about 8 meters depth near Cape MacClear, Lake Malawi. The rocky biocover is constantly being fed upon by the fishes whether it be for the algae, the fauna associated with it, or both. Note the angles at which the different fishes feed. This depends a great deal upon the type and position of the teeth and whether there is a snout as in *Labeotropheus*. Photos by Dr. Herbert R. Axelrod.

example, *Haplochromis similis* of Lake Malawi has an outer bicuspid row of teeth and inner rows of tricuspid teeth. The outer teeth are so shaped that when a plant leaf is grasped (held tightly by the tricuspid teeth) they are able to cut through it. The pharyngeal teeth break down the indigestible cellulose walls to expose the digestible portions.

Some species are more or less indiscriminate as to what they feed on—as long it is of the proper size. These are the omnivores, which utilize both animal and plant material. They can be plankton feeders feeding on the drifting algae and drifting animals in the form of small crustaceans, larvae of various sorts, eggs, etc., or feeders on the rocks which will take any animal material living among the algae along with the algae themselves. *Eretmodus cyanostictus* of Lake Tanganyika feeds in this manner, as does *Haplochromis fenestratus* of Lake Malawi. An example of an omnivorous plankton feeder is *Lamprologus moorii* of Lake Tanganyika.

Some of the plankton feeders select only zooplankton (drifting animals). In Lake Malawi there is an entire group of *Haplochromis* species adapted for this type of feeding; they have been given the special name "utaka." They are generally schooling fishes that are up in the water column selecting individual prey animals to feed upon. They have a mouth that can be protruded into a sucking tube to help capture the prey. In Lake Tanganyika *Lamprologus brichardi* and *Cyprichromis* species are some of the species that fill this particular feeding niche, and they too are schooling species. One of the Malawian mbuna, *Cynotilapia afra*, would be expected to be a rock biocover feeder. In appearance it looks very much like a dwarf *Pseudotropheus zebra*. In fact, except for size it is very difficult to distinguish the two species without examining the teeth. *Cynotilapia afra* differs from *P. zebra* in having only unicuspid teeth in its mouth. And instead of working over the rock surfaces for food, it has taken to feeding on the zooplankton in the vicinity of the rocks. It does not rise in the water column very far but can be seen up to a meter away from the rocks.

There are several species that select only small animals from the rock biocover. In many cases there have been modifications which enable them to do this quite efficiently. Most notable of these fishes are members of the genus *Labidochromis*. The head has been compressed (greatly in some species) and the teeth have become elongate and unicuspid, adapted to more easily

On the flat sandy bottom of Lake Malawi these fishes are searching for tidbits to eat. Some grab mouthfuls of sand, sift it through their gill rakers, and retain what is edible. Others select individual items. Photo by Dr. Herbert R. Axelrod.

reaching into the algal mass and into small crevices to grab the small invertebrates. The most proficient of these species are able to feed on the animals (crustaceans, aquatic insects, etc.) without ingesting too much of the algae. Aside from the Malawian *Labidochromis* species, *Paralabidochromis victoriae* from Lake Victoria and *Tanganicodus irsacae* from Lake Tanganyika may be cited as examples.

Even the organisms that live on the sandy floors are not safe. Some cichlid species are what are called sand sifters. They will take up mouthfuls of sand and expel the sand, leaving small animals behind to be swallowed. In Lake Malawi this feeding mode has been adopted by most members of the genus *Lethrinops;* in Lake Tanganyika this niche is occupied by *Xenotilapia* species, *Callochromis* species, and *Asprotilapia* among others. In *Lethrinops* the sand is actually passed through the gill arches, the gill rakers sieving out the desired material; in *Xenotilapia* the sorting is accomplished in the mouth and the sand is ejected from the mouth.

In shallow water the algal growth is quickly replenished and the grazers have a steady supply of food. Seen here are some Malawi mbuna grazing on the biocover. Photo by Dr. Herbert R. Axelrod.

In both these photos taken underwater in Lake Malawi virtually every fish is working over the biocover of the rocks. The *Labeotropheus fuelleborni* (blue fish in upper photo) has to feed at a different angle because of its protuberant snout. Upper photo by Dr. Warren E. Burgess; lower photo by Dr. Herbert R. Axelrod.

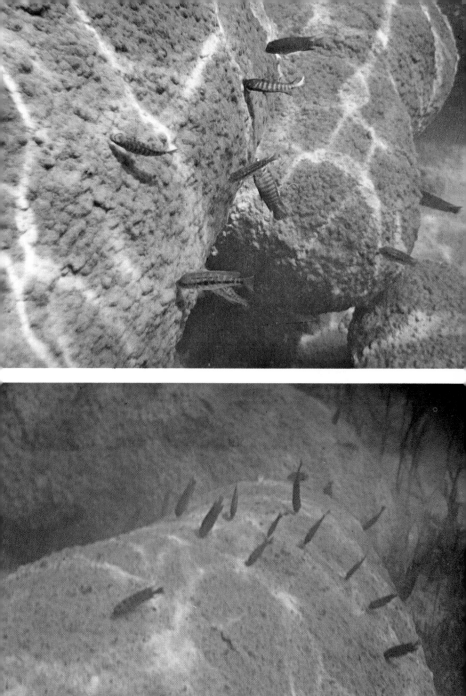

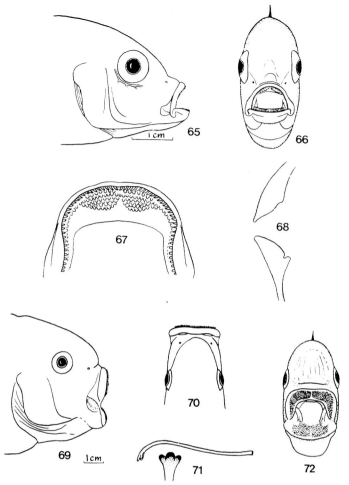

Heads, jaws and teeth of a scraper of rocks and leaves, and a rock-scraper, from L. Malawi. 65–68—*Cyathochromis obliquidens*: 65—Head, lateral; 66—Head, anterior; 67—Upper dentition; 68—Opposing outer teeth of upper and lower jaws. 69–72—*Petrotilapia tridentiger*: 69—Head, lateral; 70—Head, dorsal; 71—Single tooth seen from the side, with detail of its expanded tip in face view; 72—Head, anterior

The sketches and caption above were taken from Fryer and Iles' *"The Cichlid Fishes of the Great Lakes of Africa."*

Molluscs are an excellent food for fishes but, because of the molluscs' hard protective shells, few are able to capitalize on this source. Certain cichlids in all three lakes have modified pharyngeal teeth for just this purpose. The pharyngeal teeth are flattened or plate-like and form a surface for crushing the shells. Although some fishes have strong enough jaws (and teeth) for

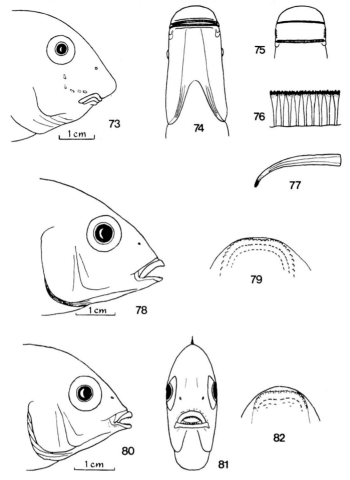

Heads, jaws and teeth of an algal-eating rock scraper, a specialist picker of algal filaments, and a general *Aufwuchs* eater all from L. Malawi; 73–77—*Labeotropheus fuelleborni*: 73—Head, lateral; 74—Heads, ventral; 75—Jaws, semi-ventral, with mouth open; 76—Anterior row of teeth, anterior; 77—Single tooth from this row, lateral. 78–79—*Haplochromis guentheri*: 78—Head, lateral; 79—Upper dentition. 80–82—*H. fenestratus*: 80—Head, lateral; 81—Head, anterior; 82—Upper dentition

The sketches and caption above were taken from Fryer and Iles' *"The Cichlid Fishes of the Great Lakes of Africa."*

this purpose, the pharyngeal teeth are more commonly modified. At least one species, *Haplochromis sauvagei* of Lake Victoria, is able to remove the snail, through leverage, from the shell without having to crush it. The mollusc-eaters with the heavy pharyngeal bones include such species as *Lamprologus tretocephalus* and *Lobochilotes labiatus* (which crushes crab carapaces as well) of

Haplochromis mloto is a member of the utaka group of *Haplochromis* species of Lake Malawi. The utaka have a protrusible mouth that can extend to form a sucking tube, an adaptation for feeding on the plankton. The upper photo shows the tube fully extended (feeding position); the lower photo shows it almost completely contracted. Upper photo by Michael K. Oliver, lower photo by Dr. Herbert R. Axelrod.

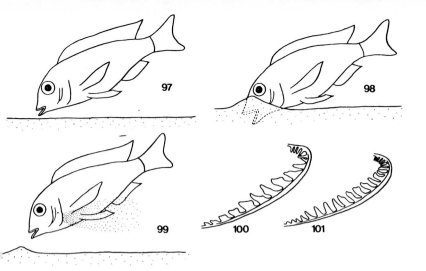

Diggers in sand from L. Malawi. 97–99—Successive positions as *Lethrinops furcifer* fills its mouth with sand and passes collected material through its gill raker sieve; 100—Outermost row of gill rakers; 101—Outermost row of gill rakers. *Lethrinops* sp. (Note the difference in spacing of the gill rakers, which is associated with selection of different foods by these closely related species.)

One of the *Haplochromis* species *(H. burtoni?*) feeding in a manner similar to that shown above. Photo by H. Hansen, Aquarium Berlin.

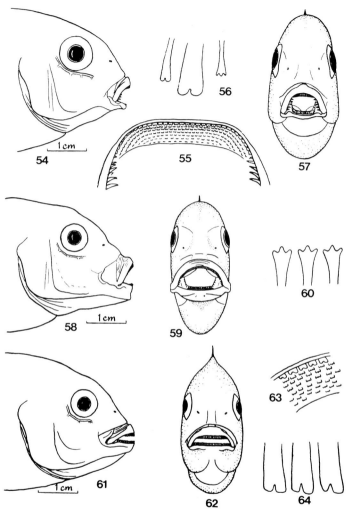

Heads, jaws and teeth of three algal-eating rock scrapers from L. Malawi. 54–57—*Pseudotropheus tropheops*: 54—Head, lateral; 55—Dentition of upper jaw; 56—Details of bicuspid and tricuspid teeth of upper jaw; 57—Head, anterior. 58–60—*P. zebra*; 58—Head, lateral; 59—Head, anterior; 60—Arrangement of tricuspid teeth of posterior rows. 61–64—*P. fuscus*: 61—Head, lateral; 62—Head anterior; 63—Arrangement of teeth in upper jaw; 64—Teeth of outer row of upper jaw

The sketches and captions above and opposite were taken from Fryer and Iles' "The Cichlid Fishes of the Great Lakes of Africa."

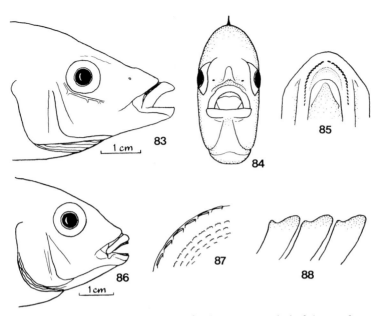

Heads, jaws and teeth of a plant scraper and a leaf chopper from L. Malawi. 83–85—*Hemitilapia oxyrhynchus*: 83—Head, lateral; 84—Head, anterior; 85—Upper dentition. 86–88—*Haplochromis similis*: 86—Head, lateral; 87—Part of upper dentition; 88—Outer teeth of upper jaw, lateral

Lake Tanganyika, *Haplochromis incola* and *H. placodon* of Lake Malawi, and *H. ishmaeli* and *Astatoreochromis alluaudi* of Lake Victoria. *Macropleurodus bicolor* of Lake Victoria and *Chilotilapia rhoadesii* of Lake Malawi are examples of cichlids that crush molluscs in their jaws.

And of course there are the cichlids that prey on other fishes. Some of these predators attack directly, using speed to capture their prey; others depend on camouflage or behavioral means to launch a "sneak attack" upon the fishes of the lakes. The direct-approach predators are often streamlined powerful swimmers that simply catch a fish too far from the rocks or other protection. *Rhamphochromis* species of Lake Malawi and *Bathybates* species of Lake Tanganyika are examples. Several laterally compressed species seem to rely on "invisibility," since looking at them head-on makes them almost impossible to see. *Lamprologus compressiceps* of Lake Tanganyika and *Haplochromis compressiceps* of Lake Malawi are examples. This latter species is also called the Malawian eye-biter because of its reputation for dining on the eyes of larger fishes. Recently, however, doubt as to whether this reputation is deserved has arisen, and aquarists

61

Lobochilotes labiatus is one of several species that have enlarged lips which are reported to be sensory in nature and used as an aid in finding prey animals. Photo by Glen S. Axelrod.

Concentrations of fishes such as seen in this photo hovering in the openness of the water column usually mean that the plankton is abundant. Photo by Dr. Herbert R. Axelrod.

Species of *Lethrinops* such as this *L. furcicauda,* are sand-diggers and have their eyes set well back from the mouth to keep them clear of the sand when feeding (see p. 59). Photo by Dr. Herbert R. Axelrod.

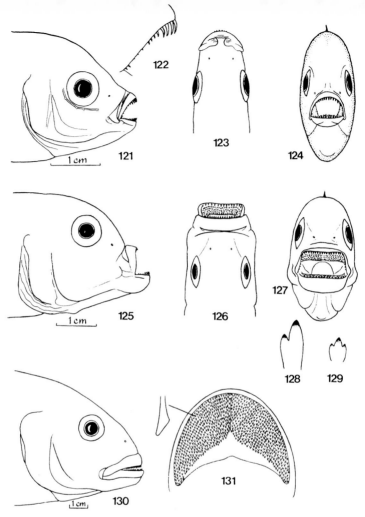

A zooplankton feeder and two scale eaters from L. Malawi. 121–124—*Cynotilapia afra*: 121—Head, lateral; 122—Teeth of upper jaw; 123 —Head, dorsal; 124—Head, anterior. 125–129—*Genyochromis mento*: 125— Head, lateral; 126—Head, dorsal; 127—Head, anterior; 128—Outer tooth of lower jaw; 129—Inner tooth of lower jaw. 130–131—*Corematodus shiranus*: 130—Head, lateral; 131—Upper dentition.

The sketches and captions above and opposite were taken from Fryer and Iles' *"The Cichlid Fishes of the Great Lakes of Africa."*

have successfully kept *H. compressiceps* in aquaria with other fishes without any problems. *H. livingstoni* of Lake Malawi lies on its side as if dead until a fish comes too close. It then suddenly comes alive to dispatch the unsuspecting fish. *H. rostratus* also has an unusual feeding method. It lies buried in the sand with only

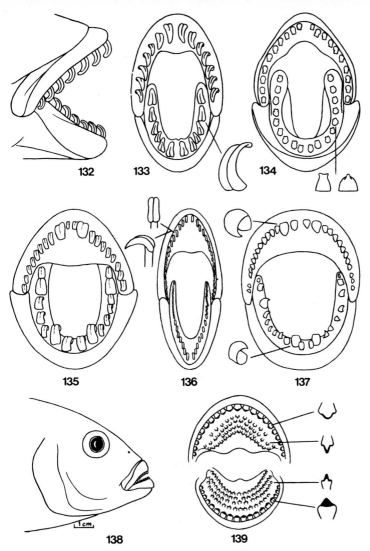

The scale eaters of L. Tanganyika and a fin biter from L. Malawi. 132 and 133—*Plecodus paradoxus*. 134—*Perissodus microlepis*. 135—*Plecodus straeleni*. 136—*Plecodus multidentatus*. 137—*Plecodus elaviae*. All show the jaws either laterally or from in front. 138 and 139—*Docimodus johnstoni*: 138—Head, lateral; 139—Dentition of the jaws. Redrawn after Boulenger and Poll.

the upper part of its head exposed. A prey fish passing by is quickly gobbled up as *rostratus* bursts from its cover. *Haplochromis johnstoni*, with vertical barring, effectively disappears from view when it hides among the vertically leaved plants like *Vallisneria*.

Of course the prey animals also have their adaptations for

Petrotilapia tridentiger of Lake Malawi feeding off the rock surface (above). It has all tricuspid teeth (lower photo) which enables it to more efficiently scrape off the algal filaments. Photos by Dr. Herbert R. Axelrod.

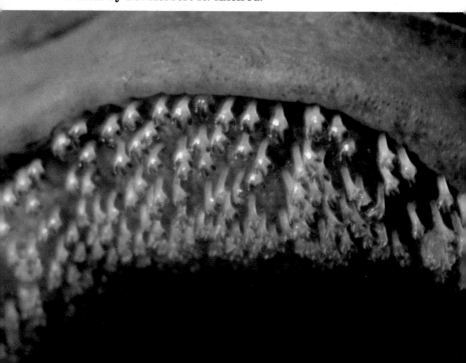

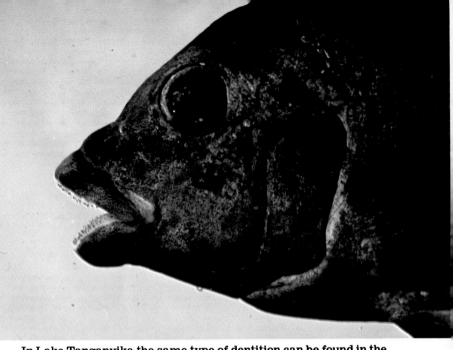

In Lake Tanganyika the same type of dentition can be found in the species of the genus *Petrochromis*. Shown here are the head (above) and mouth (below) of *Petrochromis polyodon*. Upper photo by Glen S. Axelrod; lower photo by Dr. Herbert R. Axelrod.

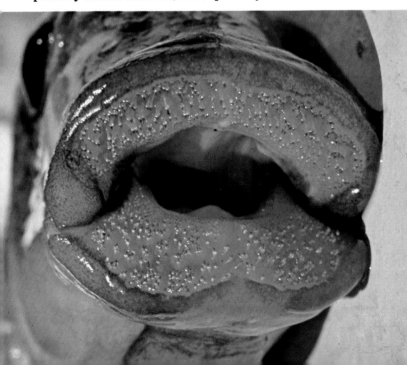

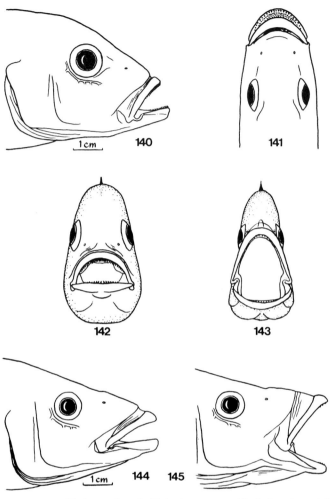

Piscivores from L. Malawi. 140–142—*Haplochromis pardalis*: 140—Head, lateral; 141—Head, dorsal; 142—Head, anterior. 143–145—*H. Polyodon*: 143—Head, anterior; 144—Head, lateral; 145—Head, lateral, with mouth open

The sketches and caption above were taken from Fryer and Iles' *"The Cichlid Fishes of the Great Lakes of Africa."*

avoiding the predators. Some are fast swimmers that rely on speed alone for safety; others are schooling species with numbers on their side (safety in numbers). Still others are camouflaged to fade into their background whether it be the sandy bottom or blue open waters.

If one remains quietly underwater the fishes, at first startled by your presence, will return to the business at hand—feeding. Photo by Dr. Herbert R. Axelrod.

Many of the fishes discussed are feeding specialists, whereas most cichlids are more generalized. Carnivores, such as many of the *Lamprologus* species, will take anything they can catch as far as animal prey is concerned. Pickers at the rock biocover will feed on what they can find and usually wind up with a combination of animal and vegetable matter. The categories of feeders are not usually very discrete, with the overlap in many instances being considerable. When these fishes are set up in aquaria they almost always will accept a wide variety of foods regardless of what specialists they were in the lakes.

A chapter on feeding cannot be complete without reference to the scale-eaters. These cichlids, although they have a highly specialized feeding mode, can be found in all three lakes. In Lake Malawi there are species of *Corematodus* as well as *Genyochromis mento*, in Lake Tanganyika the species of *Perissodus* occupy this niche, and in Lake Victoria there is a species of *Haplochromis (H. welcommei)* that is a scale-eater. Each has its adaptation to this habit, but the most peculiar seem to be the *Perissodus* species, some of which have the head and jaws skewed to one side to facilitate the scale biting but are limited to attacking from only one particular side. The scales are difficult to swallow individually and are "stacked up" in the mouth before swallowing. These fishes are, of course, not usually kept in home aquaria.

Breeding African Lake Cichlids

5

A pair of *Perissodus microlepis* guarding their young which have been released from the parent's mouth so that they could graze on the biocover. Photo by Pierre Brichard.

A pair of *Pseudotropheus tropheops*, female above and male below. The more angular anal fin of the male has two "egg spots" while that of the female has no clearly defined spots. Photo by Andre Roth.

There are basically two types of reproduction in the African rift lake cichlids—substrate spawning and mouthbrooding. Mouthbrooders are by far the most common, although the substrate spawners are not by any means rare. Most of the cichlids endemic to Lakes Malawi and Victoria are mouthbrooders, as are many of the Lake Tanganyika species. The substrate spawners of Lake Tanganyika are members of the following genera: *Lamprologus*, *Boulengerochromis*, *Telmatochromis*, *Julidochromis*, *Chalinochromis* and *Tilapia* (plus possibly some other genera in which the breeding mode is not yet known).

Of the mouthbrooders, the vast majority are maternal mouthbrooders; that is, the female parent incubates the eggs. Biparental mouthbrooding, in which both parents incubate the eggs, is known in the lakes and vicinity in such fishes as *Sarotherodon galilaea* and *Labidochromis* species. Paternal mouthbrooders (male incubates eggs) are not known from the lakes.

The number of eggs in mouthbrooders is normally considerably smaller than those of substrate spawners. For example, *Tilapia zillii* (substrate spawner) could produce as many as 7,000 or more eggs per spawn; mouthbrooders are known to produce closer to a maximum of 700, although normal average numbers would be considerably less. In some extreme cases only a half dozen or so eggs are incubated. The Lake Malawi mbuna produce from about six or so to as many as 150 eggs.

In Lake Malawi at least the reduction of the number of eggs produced by mouthbrooders seems to be correlated with a reduction in size of the left ovary (ex. *Haplochromis* spp., *Rhamphochromis* spp.) or even to having the left ovary nonfunctional (most mbuna). The eggs may be fewer in number, but they are larger and contain a large amount of yolk. Most eggs of mouthbrooders are therefore bright yellow-orange in color.

In general, mouthbrooder males develop a brighter coloration during spawning than the females. Those inhabiting sandy areas and which are normally pale in color exhibit more pastel shades, whereas the rock-inhabiting species have bolder, more vivid con-

The male *Pseudotropheus zebra* will normally select a flat rock in his territory as the breeding site and will attempt to attract females to this spot. Photo by Dr. Herbert R. Axelrod.

Lamprologus brichardi of Lake Tanganyika are normally schooling fish, but pairs break off from the school to spawn among the rocks. Photo by Dr. Herbert R. Axelrod.

In support of his theory on egg spots, Dr. W. Wickler published the above illustration. A photograph by R. Zukal (below) seems to have captured a breeding pair of the same species at the same instant.

A *Tropheus moorii* fry with the yolk still evident. At this stage the fry are still contained within the mouth of the parent. Photo by Glen S. Axelrod.

A substrate spawner of Lake Tanganyika, *Lamprologus tetracan-thus,* guarding its fry under the shelter of a rock. Photo by Glen S. Axelrod.

trasting colors, particularly blue. Blue is very prominent in male lake cichlids and in some species is normal to both sexes (ex. *Labidochromis caeruleus, Haplochromis moorii,* the blue of the male moorii intensifying during spawning). This is why it is so unusual to find such a species as *Pseudotropheus lombardoi* in which the female is bright blue and the male yellow.

The brightly colored males also tend to set up territories, whether it be a parcel of sand, a rock, cave, etc. The prominent colors of the male, which are displayed to the utmost by spread fins, movements, etc., will advertise his presence to available females and serve as a warning to all trespassers to keep away. Those which do not heed the warning will be met with a challenge or fight if necessary. *Pseudotropheus tropheops* defends its territory, often against large predatory *Haplochromis* species; *Haplochromis eucinostomus,* on the other hand, will chase off only smaller fishes, allowing larger species like *H. rostratus* to pass through even at the expense of a partially destroyed nest.

Within the guarded territory the male will usually construct a nest or at least have an area cleaned where eggs may eventually be deposited. *Haplochromis heterodon* of Lake Malawi builds a saucer-shaped nest in the sand by moving the sand a mouthful at a time to the "correct" position. *Cyathopharynx furcifer* of Lake Tanganyika builds sand nests on the tops of high boulders. It first cleans the nest site, then carries sand up from the base of the rocks to construct the nest. The highest point seems to be the most desirable, and males compete for these choice spots. The reason for the selection of the high points seems to be the mid-water position of the females. The highest nests are closest to the females, so the males in those nests are the most successful in attracting them for spawning. *Haplochromis quadrimaculatus,* a member of the open water utaka, migrates inshore to breed off the more secure rocky shores.

The males then, by their most attractive colors and by their actions, attract suitable females (egg-laden and ready to spawn) to the prepared nest. The male will normally swim toward the female and present his side (giving her the best view of his magnificent colors), then swim back to the nest. When a female follows him to the nest site his courting actions intensify, including in many species some nudging of the vent area, some circling maneuvers, etc. The eggs are laid on the substrate and either fertilized there and then picked up by the female or first picked up by the female and then fertilized by the male. A great

When Dr. E. Trewavas divided the genus *Tilapia* into *Tilapia* and *Sarotherodon,* the old familiar *Tilapia mossambica* became *Sarotherodon mossambicus.* It is a mouthbrooder in which the females take over the chores of brooding the eggs and fry. Species of the genus *Tilapia* are all substrate spawners; those of *Sarotherodon* are all mouthbrooders. Photo by G. Marcuse.

The female *Julidochromis ornatus* joins the male in preparing the spawning site. Photo by H.-J. Richter.

As the male waits, the female *Julidochromis ornatus* lays a few eggs (not visible). Photo by H.-J. Richter.

The male *Julidochromis ornatus* quickly fertilizes the pair of eggs laid by the female. Photo by H.-J. Richter.

After spawning is completed the eggs are fanned and guarded by the female as they develop. Photo by H.-J. Richter.

deal of controversy clouds these actions, but with more and more observations of the spawning habits of these fishes both in aquaria and in the lakes themselves, the pattern is beginning to become clearer.

At the end of the spawning the female is left with a mouthful of eggs which she will incubate until the fry are relatively independent. In some species the female will remain in the male's territory, protected from trespassers by the guarding male; in other species the male will drive the female away, or she will leave the site on her own to search out some secluded place in which to brood the eggs and fry. Normally the female will not feed during this incubation period for obvious reasons, but some species, such as *Tropheus* spp., may take some nourishment while brooding. During the incubation period the water drawn through the mouth and gill area flows over the eggs and fry, providing the means for the gaseous exchange. In addition the females can be seen "chewing" the eggs. This chewing motion actually is a shifting of the eggs in the mouth, the change in position preventing the yolk of the egg from settling at one pole and thereby causing a retardation of development.

The length of incubation time (from spawning to first release of the fry) varies from species to species and in addition depends upon the temperature. Average mbuna incubation is about 22 to 30 days, whereas *Tropheus* species are said to take up to 42 days, with a week or two more for the care of the fry. Most mouthbrooders will tend the fry for a period after the first release, but some species (ex. *Labidochromis vellicans*) will release them and then ignore them. The first food of the fry in the lakes appears to be small particulate matter including diatoms and tiny algae.

Various species brood their eggs and fry in various places. As mentioned, some females remain in the male's territory, whereas others seek refuge among the nearby rocks. Some offshore species, such as the Malawian *Sarotherodon,* will come into shallow reedy areas to brood. Similarly the deep-dwelling *Lobochilotes labiatus* and *Perissodus* species will come up to shallower water to brood the young.

Certain sand-nest spawners, such as *Xenotilapia* species, spawn simultaneously, apparently reacting to group stimuli. The females swimming overhead are attracted to a particular nest, swim down and spawn in it, and resume their mid-water position. Spawners among the rocks include the mbuna of Lake Malawi

A pair of *Sarotherodon mossambicus,* the more brightly colored male below the female. Photo by Milan Chvojka.

A male *S. mossambicus* with the lips characteristic of an older fish. Photo by Gerhard Marcuse.

A pair of *Lamprologus brichardi* spawning in an aquarium. It is difficult in this species to determine which is the male or female until actual spawning begins. Photo by H.-J. Richter.

The site selected is usually a smooth rock such as those in the background. Here courting continues. Photo by H.-J. Richter.

plus *Cyphotilapia frontosa* (deep water), the goby cichlids (shallow surf-washed shorelines), *Simochromis*, *Tropheus*, and *Petrochromis*, all of Lake Tanganyika.

Part of the color pattern of many of the mouthbrooders is a group of spots in the anal fin. These have been called "egg spots" because their size, shape, and color are similar to the eggs that are laid by these species. In species like *Haplochromis burtoni* and its close relatives the resemblance is quite close, and a theory was proposed by Wickler that they are actually egg dummies. In these (and some other species) the eggs are said to be laid by the female in the normal way but taken up by the female so quickly that the male will have had no chance to fertilize them. The male, however, extends the anal fin in a flat position on the nest site, "displaying" the egg dummies which the female takes for the real thing and attempts to add them to the ones already in her mouth. By her attempt to pick up the false eggs she must approach the vent of the male. As he is extruding sperm the female

As the pair of *Sarotherodon mossambicus* prepare to spawn, they dig a deep hole in the sand. Photo by G. Marcuse.

As soon as the male fertilizes the eggs, the female takes them into her mouth. In this species the eggs are fertilized before the female picks them up. Photo by G. Marcuse.

A male S. *mossambicus* that looks like he is carrying eggs. Either this is a false impression or it is a rare instance where a male is brooding eggs. Photo by G. Marcuse.

Courtship continues as the male attempts to entice the female back to the spawning site. Note the eggs already deposited. Photo by H.-J. Richter.

The courtship can become quite acrobatic as the two fish react to each other's behavior. Photo by H.-J. Richter.

A male *Sarotherodon mossambicus* with some fry. The fry are feeding on some microorganisms or algae which are present on the log. Photo by Gerhard Marcuse.

in her pecking and breathing will draw the sperm in and the eggs become fertilized. In *Pseudocrenilabrus* species the actions are similar but delayed to some extent; it is theorized that the eggs are fertilized before and after the picking up by the female. Most species, however, have small egg spots or none at all. Where they are present they may be only a sort of recognition device or perhaps a sign stimulus to elicit the behavior in the female of searching for the eggs. In some species the egg spots are unusually situated. For example, in a species of *Petrochromis* they are on the membrane flaps at the tip of the anal fin spines. In other species, ex. *Cyathopharynx furcifer, Ophthalmotilapia* spp., etc., the egg spots are located on the tips of the elongate rays of the ventral fins. When folded back this would bring the egg spots into the proper position along the anal fin, but observations by Pierre Brichard on the breeding mode of these species indicate that they do not breed in the same way as the other mouthbrooders. In fact the females lay their eggs in mid-water and have them in their mouth before approaching the nest. The male has already deposited his sperm and left before she arrives. What may happen is that the females will follow the bright egg spots of the male

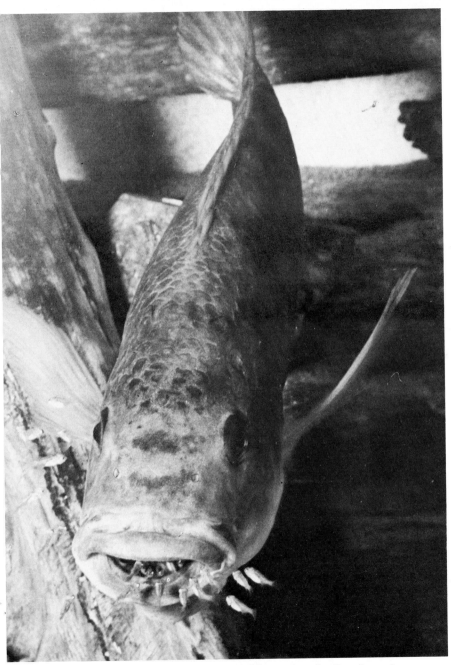

The female *Sarotherodon mossambicus* signals the fry when
danger threatens. They swiftly return to the refuge of her mouth.
Photo by Gerhard Marcuse.

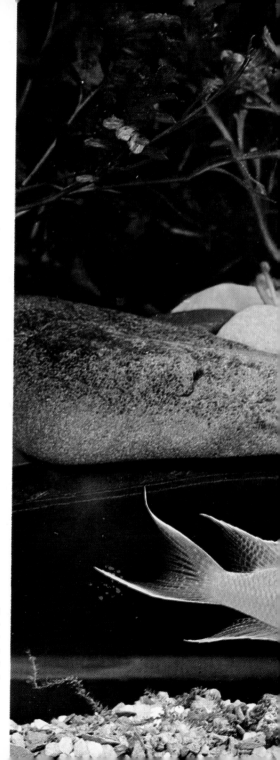

Lamprologus brichardi approaching the spawning rock in preparation for the depositing of more eggs. Photo by H.-J. Richter.

90

A male *Sarotherodon mossambicus* (above) with his distinctive white chin guarding the free-swimming fry. The female (below) is releasing the fry so that they may feed. Photos by Gerhard Marcuse.

In *Sarotherodon macrochir,* the female mouths the genital tassle of the male. Her mouth is swollen with eggs. The thin string is a sperm thread and will facilitate fertilization of the eggs in the female's buccal cavity. Photo from Fryer and Iles' *The Cichlid Fishes of the Great Lakes of Africa.*

to the nest site and that the spots are useful only as an attracting device and guide so that the female will arrive at the proper spot where the water is laden with sperm.

The substrate spawners will also prepare nests in the sand or among the rocks, but in many cases the nest preparation will be a joint affair with both sexes participating. Where the mouthbrooders come together just for spawning, the substrate spawners often will form pairs for longer periods of time—at least until the fry are somewhat able to fend for themselves. When territories are defended both sexes may also participate, although more often than not it is the male that has the greater burden. Courting is similar, with the males generally the more colorful and active to attract a mate. The eggs are laid on the substrate, usually in a precise manner in adjacent strings, their position maintained by their stickiness. The male follows immediately to fertilize them. The parents, usually also as a joint venture, will tend them, guarding them from predators, fanning them to promote adequate gaseous exchange, and removing the dead eggs.

Labeotropheus fuelleborni fry taken from the mouth of the female parent. They are almost at the size where they would be released anyway. Photo by Dr. D. Terver, Nancy Aquarium, France.

The female *Labeotropheus trewavasae* carrying her brood. The characteristic distension of the gular (throat) area is evident. Photo by H. Hansen, Aquarium Berlin.

A female *Pseudotropheus* sp. In the upper photo she is picking up some eggs from the aquarium gravel; in the lower photo she has finished as indicated by her extended buccal area. For safety she can be removed from the community aquarium to incubate the brood in her own private quarters.

The fry of *Sarotherodon mossambicus* are let out of the parent's mouth to feed. In the upper photo they are milling about. But at a signal from the mother they will retreat to the safety of the mouth. In the lower photo they have received the signal and are all intent on reaching safety. Photos by G. Marcuse.

It is almost impossible to believe that these photos were taken within ten seconds of each other!!! The female has almost all her babies in safely. Photo by Gerhard Marcuse.

Hatching occurs in a matter of a few days; the yolk is generally absorbed in another five or six days. The fry (and sometimes the eggs) may be moved about from one place to another (in sand spawners from one pit to another) perhaps for protection or cleanliness or both. The fry are also guarded by the parents, the female more usually staying closer to the fry, the male patrolling the outer perimeter of their territory. *Boulengerochromis microlepis* comes up from deep water to spawn in the shallows, where it constructs a large crater-like nest in the sand. This large species may deposit as many as 12,000 to 15,000 eggs in this nest. Other species prefer the shelter of an overhanging rock or a cave (ex. *Telmatochromis caninus, Julidochromis* spp., several *Lamprologus* species, etc.). Some small species of *Lamprologus (L. brevis, L. ocellatus,* and *L. multifasciatus)* stake out a snail's shell (usually *Neothauma*) as their territory and nesting site. *Lamprologus brichardi* is normally a schooling species, but spawning is an individual matter. Pairs will separate from the school, and the eggs are laïd in nests among the rubble.

97

Breeding African Cichlids in the Aquarium

6

A spawning pair of one of the *Pseudotropheus* species. The female has the less brilliant color and can be seen to have at least some eggs in her mouth already. Photo by Aaron Norman.

Pseudocrenilabrus philander spawning in an aquarium. Note the female with a mouth full of eggs attracted by the egg dummies in the anal fin of the male who displays them over the nest. Photo by Ruda Zukal.

It is obvious that having extensive areas in which huge sand nests can be built or having large boulders among which the fishes can flit for protection is impossible in the home aquarium. But the cichlids are so adaptable and accommodating that they will readily spawn (at least many of them) under the conditions we set up to try and re-create their natural surroundings. Needless to say, large aquaria are the rule, with 200-liter tanks almost too small. Of course certain species will do comfortably in smaller tanks, but for most species the large aquaria are a must.

For those cichlids that build a sand nest it is suggested that the aquarium have about four or more inches of coarse sand in the bottom. A few rocks or artificial plants might be helpful for decoration as well as providing some refuge for a beleaguered female or subordinate male. For mbunas and other rock-dwellers a tank set up with a lot of rockwork forming many caves is necessary. But a word of caution here—the mbunas are diggers and will soon excavate all the sand or gravel out from under the rocks (and usually place it in front of the front glass of the aquarium—their own little attempt at privacy!). If the rocks are poorly placed they will collapse, possibly injuring the fishes or even cracking the tank itself. Be sure to set up the rocks as firmly as possible. This can be accomplished by placing them in the aquarium before the sand so that they sit on the bottom of the tank itself, not on a base of sand. For substrate spawners, and even for the mouthbrooders, provide a choice of spawning sites such as a flat rock in a secluded spot or a cave which has a flat area for spawning (including a flat ceiling for such species as *Julidochromis ornatus,* etc.).

Since many of the African cichlids, especially dominant males, are territorial, it is often better to have only one male of a species in a tank unless that tank is so large that several territories can be set up without overlapping. Unfortunately there are many fishes that will consider the entire tank as their territory and try to evict all inhabitants. If a new fish is introduced into a tank in which territories have already been established there likely will be serious

This is a typical rocky outcropping in Lake Malawi. The rocks continue under the water in about the same configuration. Remember this when setting up an African cichlid tank. Photo by Dr. Herbert R. Axelrod.

trouble. A safe technique for the introduction of new fishes to an established aquarium is to shift the rocks around so that all fishes have an equal chance to establish new territories. Some pet shop owners keep their African cichlids in tanks in which the fishes have no place to hide. This prevents them from setting up territories and provides for a better display of the fishes offered for sale. Unfortunately, under such conditions the fishes' colors soon fade, and if one fish does decide to attack another the loser cannot easily get out of its way.

There is one other consideration in the tank with many rocks. You might find it necessary to remove a female holding eggs in her mouth or to take out a sick fish, a bully, or an injured fish. It is therefore advisable that when you set up the rocks you do it with the understanding that you may be chasing a fish around them or removing some of them in the process. Make it easy on yourself (but not to the extent that the fishes will suffer).

The fry above were taken from the buccal cavity of a female *Melanochromis auratus* a few days after the eggs hatched. The lower photo shows the fry a few days later, having absorbed much of their yolk and almost ready to look out for themselves. Photos by Gerhard Marcuse.

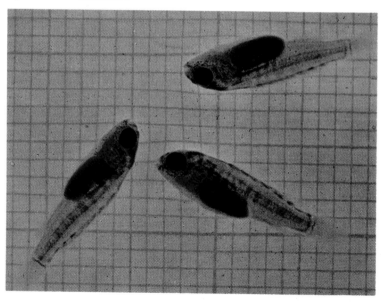

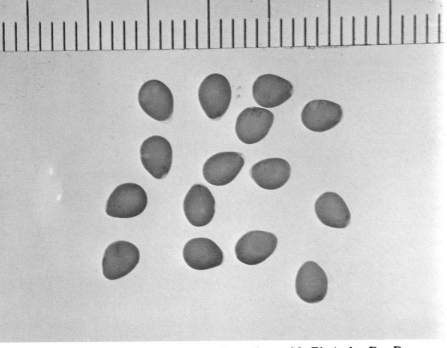

Melanochromis auratus eggs, three days old. Photo by Dr. D. Terver, Nancy Aquarium, France.

A female *Melanochromis auratus* releasing her young to feed. Note that the young appear like miniature female adults, the distinctive male coloration not appearing until they reach maturity. Photo by A. F. Orsini.

A rocky area of Lake Tanganyika. Most of the fishes visible in this photo are *Lamprologus brichardi*. Photo by Dr. Herbert R. Axelrod.

Opposite above: Mouthbrooder fry are released at the base of the rocks where they are relatively safe from the fishes above. These rocks are at a depth of about 10 meters in Lake Malawi. Photo by Dr. Herbert R. Axelrod. *Below:* Some cichlids prefer grassy areas or sandy areas to the rocks for breeding and tending the eggs and fry. A grassy habitat in Lake Malawi. Note the barred predatory *Haplochromis*. Photo by Dr. Herbert R. Axelrod.

Mbunas and Tanganyikan rock-dwelling cichlids do very well in outdoor cement pools in Florida. The high alkalinity suits them quite well, since a pH of 7.4 or higher is usually recommended for these fishes. All that has to be done is to add a mixture of sexes to the pool and they will breed on their own. Much the same will happen in your tanks with the mouthbrooders. Just add a male and two or more females to your tank and wait. Eventually, with good feeding and clean water, you will find the female with eggs in her mouth. After a day or two remove her to a separate tank with some rocks for shelter and keep her there until the fry are released, about three weeks' time. It's that simple.

The substrate spawners are more difficult but spawn much like their South American cousins. Similar treatment and feeding is all that is necessary. As long as the water is hard and alkaline and of the correct temperature they will probably spawn for you.

Diseases and Parasites

A *Haplochromis moorii* with a disease (possibly myxosporidiasis) affecting the caudal peduncle. Photo by H.-J. Richter.

African rift lake cichlids can be quite rough in an aquarium, leaving some individuals in a sorry state. If the fish (in this case *Haplochromis longimanus*) doesn't die from its wounds it will be easy prey for any infection. Always provide plenty of cover to reduce losses. Photo by Dr. Herbert R. Axelrod.

African lake cichlids are relatively hardy fishes and, although they are susceptible to almost all of the diseases of aquarium fishes, they rarely develop them. But even in the wild there is an occasional cichlid that shows the symptoms of bacterial and fungal infections. The fungal infections seem to attack fishes that have been damaged in some way (by an unsuccessful predator perhaps), cases in which the protective covering of scales and mucus has been lost. Damage of any kind in the aquarium should be treated promptly to prevent fungal infections.

Ichthyophthiriasis has also been observed on cichlids in the lakes; this disease is probably the cause of some mortality in young or weak fishes.

A dangerous disease called the Malawi bloat has turned up in African cichlid tanks. It at first caused quite a bit of consternation with cichlid enthusiasts until remedies were developed. The fish will "bloat" up, lose its equilibrium, and eventually die unless it receives proper treatment. Several remedies are now available and are usually sold where African lake cichlids are sold.

In the thousands of lake cichlids imported a very high percentage are parasitized by digenetic trematodes. It appears that the metacercarial stage, one of the earlier stages in the life history of the parasite, bores its way into the skin of the fish and forms a cyst. The cysts are darkly pigmented and stand out noticeably on light-colored fishes. These cysts do not appear to do the fishes much harm, and other fishes in the tank are not affected. The fish is probably host for only one of the stages in the life cycle of the parasite. Without the other hosts the parasites cannot complete their life cycle and eventually die. In nature the fish might be eaten by a piscivorous bird in which the next stage of the parasite can develop.

The usual complement of other internal and external parasites such as acanthocephalans or copepods infect the African cichlids. In most cases there is no problem unless the individual is so weakened by the parasite that the travelling or handling causes it to expire. In most of these instances, however, the fish dies before

A male *Genyochromis mento* and some of its victims. Even individuals of *Genyochromis mento* are not immune from the attacks of their own species as this photo shows. Photo by A. Ivanoff.

it can make it to the retailer's tanks or is so obviously covered with parasites that nobody in his right mind would buy it.

There are many good books on the diseases of aquarium fishes, and they should be consulted for most of the problems you might encounter with your African lake cichlids. With all the newer remedies on the shelves and the hardiness of the African cichlids themselves there should be no reason why your fishes should not live long and spawn readily.

Identifying African Lake Cichlids

8

Pseudotropheus species (possibly
Pseudotropheus brevis, which
formerly was known as
Melanochromis brevis). Photo by
Andre Roth.

A typical adult male *Melanochromis
vermivorus* with the reverse pattern
of the female. *Melanochromis*
species are very poorly known.
Photo by Andre Roth.

The identification of African lake cichlids is often a difficult task. Even scientists, with many preserved specimens in front of them run into difficulties when trying to determine the correct identity of some species. They have a tremendous advantage over aquarists however, in that they can examine the different morphological characters that make such identifications possible. They can dissect out the pharyngeal teeth, count scales or fin rays, examine the teeth in the jaws, etc., comparing these features with already published descriptions of known species.

Aquarists must rely on features they can see in the living fishes. They deal with general shape, color, and pattern for their identification, comparing these with published photographs and hoping that the identification of the fish in the photo is correct. Rarely is the color of the living fish provided in the scientific description, so a comparison is virtually impossible.

It was with these thoughts in mind that this book was conceived. Photos were basically of fishes that were tranquilized and later preserved so that a positive identification could be made. The addition of aquarium photos of the species, many in their dominant or breeding patterns, made comparison with the living individuals in aquarists' tanks easier. Whenever new lake cichlids are brought to our attention they are photographed and added to the book in the next possible edition. In this way this book is kept as up to date as possible.

Unfortunately, many of the "new" fishes are not so easily identified, and in many cases somebody along the line makes up a name to fill the gap where there are no names immediately available. This leads to confusion, as the non-scientific name eventually has to be discarded in favor of the correct name. In the next two chapters current lists of the scientific names of the cichlids from Lakes Malawi and Tanganyika are included.

When trying to identify a living fish, look at the general shape first. Many of the species have features that will either serve to identify it immediately or at least narrow the choice down to only a few species. Such things as expanded lips, compressed bodies, enlarged pore areas on the head, produced snouts, etc., fall into

112

An underwater scene in Lake Tanganyika showing a school of *Lamprologus brichardi* and some striped fish, probably *Telmatochromis bifrenatus* or *T. vittatus*. These species are involved in taxonomic problems. Photo by Dr. Herbert R. Axelrod.

this category. *Haplochromis* species are generally difficult to identify. In non-breeding or nondominant individuals there is usually a pattern of dark markings that can be used to narrow the choices of what species it might be. These markings may be large blotches (as in *H. livingstoni, H. venustus,* etc.), a slanting band (as in *H. incola, H. epichorialis*), horizontal band(s) (as in *H. similis, H. strigatus,* etc.), rows of spots (several in *H. spilostichus,* few in *H. nitidus,* etc.), or even a checkered pattern as in *H. fenestratus.* Vertical bars are only sometimes useful (as in *H. electra* or *H. johnstoni*), since they are mostly transient markings depending upon mood.

Cichlids of Lake Tanganyika

9

A variety of *Julidochromis transcriptus* that has been known in the trade in some areas as *J.* "kissi", a non-valid name. Photo by Pamela and Jorgen Hansen.

One of the more popular fishes of Lake Tanganyika is this *Lamprologus leleupi*. This species has been broken down into several subspecies. Photo by H.-J. Richter.

The following list includes all the species of cichlids from Lake Tanganyika that are presently known. Since the time of Prof. Max Poll's comprehensive treatment of the lake fishes in 1956, a number of taxonomic changes have been made and more than a dozen new species have been discovered. These have been added to Poll's basic list along with indications of what new species are being studied.

Numerous recent changes have been made in genera. *Haplochromis* has been partially restructured to include its species in *Astatotilapia, Gnathochromis* (in part), and *Astatoreochromis. Limnochromis* formerly included the species here listed as *Gnathochromis* (in part) and *Lepidochromis. Lepidolamprologus* is a partitioning of *Lamprologus*, while *Ophthalmochromis* is now included in *Ophthalmotilapia.* Only one species remains in *Tilapia,* the others being transferred to *Sarotherodon.* Obviously some of these changes are controversial and will probably not be accepted, while there are sure to be other reassignments of species from such large and possibly composite genera as *Haplochromis* and *Lamprologus.* Because of the instability of these changes, we have not modified the captions to reflect the new generic status of the species in question.

For a more comprehensive study of the fishes of Lake Tanganyika see Pierre Brichard's book *The Fishes of Lake Tanganyika.*

Asprotilapia leptura Boulenger
Astatoreochromis straeleni (Poll)
Astatoreochromis vanderhorsti (Greenwood)
Astatotilapia bloyeti (Sauvage)
Astatotilapia burtoni (Günther)
Astatotilapia paludinosa Greenwood
Astatotilapia stappersi (Poll)
Aulonocranus dewindti (Boulenger)
Bathybates fasciatus Boulenger
Bathybates ferox Boulenger
Bathybates graueri Steindachner
Bathybates horni Steindachner
Bathybates leo Poll
Bathybates minor Boulenger
Bathybates vittatus Boulenger
Boulengerochromis microlepis (Boulenger)
Callochromis macrops macrops (Boulenger)
Callochromis macrops melanostigma (Boulenger)
Callochromis pleurospilus (Boulenger)
Chalinochromis brichardi Poll
Chalinochromis n. sp.
Cunningtonia longiventralis Boulenger

Cyathopharynx furcifer (Boulenger)
Cyathopharynx schoutedeni (Poll)
Cyphotilapia frontosa (Boulenger)
Cyprichromis brieni Poll
Cyprichromis leptosoma (Boulenger)
Cyprichromis microlepidotus (Poll)
Cyprichromis nigripinnis (Boulenger)
Ectodus descampsi Boulenger
Eretmodus cyanostictus Boulenger
Gnathochromis permaxillaris (David)
Gnathochromis pfefferi (Boulenger)
Grammatotria lemairei Boulenger
Haplochromis benthicola Matthes
Haplochromis horei (Günther)
Haplotaxodon microlepis Boulenger
Haplotaxodon tricoti Poll
Hemibates stenosoma (Boulenger)
Julidochromis dickfeldi Staeck
Julidochromis marlieri Poll
Julidochromis ornatus Boulenger
Julidochromis regani Poll
Julidochromis transcriptus Matthes
Lamprologus brevis Boulenger
Lamprologus brichardi Poll
Lamprologus buescheri Staeck
Lamprologus callipterus Boulenger
Lamprologus calvus Poll
Lamprologus caudopunctatus Poll
Lamprologus christyi Trewavas

and Poll
Lamprologus compressiceps Boulenger
Lamprologus fasciatus Boulenger
Lamprologus furcifer Boulenger
Lamprologus hecqui Boulenger
Lamprologus kambaensis Staeck
Lamprologus kungweensis Poll
Lamprologus leleupi leleupi Poll
Lamprologus leleupi longior Staeck
Lamprologus leleupi melas Matthes
Lamprologus leloupi Poll
Lamprologus lemairei Boulenger
Lamprologus meeli Poll
Lamprologus modestus Boulenger
Lamprologus mondabu Boulenger
Lamprologus moorei Boulenger
Lamprologus multifasciatus Boulenger
Lamprologus mustax Poll
Lamprologus niger Poll
Lamprologus nkambae Staeck
Lamprologus obscurus Poll
Lamprologus ocellatus Steindachner
Lamprologus ornatipinnis Poll
Lamprologus petricola Poll
Lamprologus prochilus Bailey and Stewart
Lamprologus pulcher Poll
Lamprologus savoryi Poll
Lamprologus schreyeni Poll
Lamprologus sexfasciatus Trewavas and Poll
Lamprologus signatus Poll
Lamprologus stappersi Pellegrin
Lamprologus tetracanthus Boulenger
Lamprologus toae Poll
Lamprologus tretocephalus Boulenger
Lamprologus wauthioni Poll
Lepidochromis bellcrossi (Poll)
Lepidochromis christyi (Trewavas)
Lepidolamprologus attenuatus (Steindachner)
Lepidolamprologus cunningtoni (Boulenger)
Lepidolamprologus elongatus (Boulenger)
Lepidolamprologus kendalli (Poll)
Lepidolamprologus pleuromaculatus (Trewavas & Poll)

Lepidolamprologus profundicola (Poll)
Lestradea perspicax perspicax Poll
Lestradea perspicax stappersi Poll
Limnochromis abeelei Poll
Limnochromis auritus (Boulenger)
Limnochromis staneri Poll
Limnotilapia dardennei (Boulenger)
Limnotilapia loocki Poll
Limnotilapia trematocephala (Boulenger)
Lobochilotes labiatus Boulenger
Ophthalmotilapia boops (Boulenger)
Ophthalmotilapia nasutus (Poll and Matthes)
Ophthalmotilapia ventralis ventralis (Boulenger)
Ophthalmotilapia ventralis heterodontus (Poll and Matthes)
Orthochromis malagaraziensis (David)
Perissodus eccentricus Liem and Stewart
Perissodus elaviae (Poll)
Perissodus hecqui (Boulenger)
Perissodus microlepis Boulenger
Perissodus multidentatus (Poll)
Perissodus paradoxus (Boulenger)
Perissodus straeleni (Poll)
Petrochromis famula Matthes and Trewavas
Petrochromis fasciolatus Boulenger
Petrochromis macrognathus Yamaoka
Petrochromis orthognathus Matthes
Petrochromis polyodon Boulenger
Petrochromis trewavasae Poll
Petrochromis sp.
Pseudocrenilabrus philander (Weber)
Reganochromis calliurum (Boulenger)
Reganochromis centropomoides Bailey and Stewart
Reganochromis sp.
Sarotherodon karomo (Poll)
Sarotherodon nilotica (Linnaeus)
Sarotherodon tanganicae (Günther)
Simochromis babaulti Pellegrin
Simochromis curvifrons Poll
Simochromis diagramma Günther

A young *Tropheus moorii* from Lake Tanganyika commonly called the "rainbow moorii" or "rainbow *tropheus*." This genus contains few species but many morphs. Photo by Pierre Brichard.

The striking white spots on the dark background make the young
Tropheus duboisi relatively easy to identify underwater, even
from a distance. Photo by Glen S. Axelrod.

One of the sandy bottom dwellers, *Xenotilapia sima* is large (to 150 mm) and not very colorful. Photo by Pierre Brichard.

One of the popular species of the large genus *Lamprologus, L. tretocephalus. Lamprologus* is the largest genus of cichlids in Lake Tanganyika, with over 35 species.

Simochromis margaretae Axelrod and Harrison
Simochromis marginatus Poll
Simochromis pleurospilus Nelissen
Spathodus erythrodon Boulenger
Spathodus marlieri Poll
Tangachromis dhanisi (Poll)
Tanganicodus irsacae Poll
Telmatochromis bifrenatus Myers
Telmatochromis burgeoni Poll
Telmatochromis caninus Poll
Telmatochromis temporalis Boulenger
Telmatochromis vittatus Boulenger
Tilapia rendalli Dumeril
Trematocara caparti Poll
Trematocara kufferathi Poll
Trematocara macrostoma Poll
Trematocara marginatum Boulenger
Trematocara nigrifrons Boulenger
Trematocara stigmaticum Poll
Trematocara unimaculatum Boulenger
Trematocara variabile Poll
Triglachromis otostigma (Regan)

Tropheus annectans Boulenger
Tropheus brichardi Nelissen and Thys
Tropheus duboisi Marlier
Tropheus moorei moorei Boulenger
Tropheus moorei kasabae Nelissen
Tropheus polli G. Axelrod
Tylochromis polylepis (Boulenger)
Xenotilapia boulengeri (Poll)
Xenotilapia caudafasciata Poll
Xenotilapia lestradei Poll
Xenotilapia longispinnis longispinnis Poll
Xenotilapia longispinnis burtoni Poll
Xenotilapia melanogenys (Boulenger)
Xenotilapia nigrolabiata Poll
Xenotilapia ochrogenys ochrogenys (Boulenger)
Xenotilapia ochrogenys bathyphilus Poll
Xenotilapia ornatipinnis Boulenger
Xenotilapia sima Boulenger
Xenotilapia spilopterus Poll and Stewart
Xenotilapia tenuidentata Poll

*There is a possibility of new species in the genera *Lamprologus, Julidochromis, Tropheus* and *Simochromis*.

Cyphotilapia frontosa is still, after many years, one of the more popular Lake Tanganyikan fishes even though the price remains relatively high. Photo by G. Budich.

Two male *Lamprologus attenuatus* fighting. One apparently resents the intrusion of the other into its territory (the rocky cave). Photo by Dr. R. J. Goldstein.

A pair of *Julidochromis marlieri*. These are substrate spawners which require secluded areas (such as caves) in which to spawn. Photo by W. Hoppe.

Cichlids of Lake Malawi

10

Haplochromis nitidus is well marked with three dark lateral spots. Photo by Dr. Herbert R. Axelrod.

Head study of *Haplochromis placodon.* Photo by Dr. Herbert R. Axelrod.

The following list includes all the species of cichlids from Lake Malawi that are presently known. The study of these fishes is continuing and there will undoubtedly be new species described from time to time.

Aristochromis christyi Trewavas
Aulonocara macrochir Trewavas
Aulonocara nyassae Regan
Aulonocara rostrata Trewavas
Chilotilapia rhoadesii (Boulenger)
Cleithrochromis bowleyi Eccles
Corematodus shiranus Boulenger
Corematodus taeniatus Trewavas
Cyathochromis obliquidens Trewavas
Cynotilapia afra (Guenther)
Cynotilapia axelrodi Burgess
Diplotaxodon argenteus Trewavas
Diplotaxodon ecclesi Burgess &
  Axelrod
Docimodus evelynae Eccles & Lewis
Docimodus johnstonii Boulenger
Genyochromis mento Trewavas
Gephyrochromis lawsi Fryer
Gephyrochromis moorii Boulenger
Haplochromis ahli Trewavas
Haplochromis anaphyrmus Burgess
  & Axelrod
Haplochromis annectens Regan
Haplochromis argyrosoma Regan
Haplochromis atritaeniatus Regan
Haplochromis auromarginatus
  (Boulenger)
Haplochromis balteatus Trewavas
Haplochromis boadzulu Iles
Haplochromis borleyi Iles
Haplochromis breviceps Regan
Haplochromis caeruleus Boulenger
Haplochromis callipterus (Guenther)
Haplochromis chrysogaster
  Trewavas
Haplochromis chrysonotus
  (Boulenger)
Haplochromis compressiceps
  (Boulenger)
Haplochromis cyaneus Trewavas
Haplochromis decorus Trewavas

Haplochromis dimidiatus (Guenther)
Haplochromis electra Burgess
Haplochromis epichorialis Trewavas
Haplochromis ericotaenia Regan
Haplochromis euchilus Trewavas
Haplochromis eucinostomus Regan
Haplochromis fenestratus Trewavas
Haplochromis festivus Trewavas
Haplochromis flavimanus Iles
Haplochromis formosus Trewavas
Haplochromis fuscotaeniatus Regan
Haplochromis gracilis Trewavas
Haplochromis guentheri Regan
Haplochromis hennydaviesae
  Burgess & Axelrod
Haplochromis heterodon Trewavas
Haplochromis heterotaenia Trewavas
Haplochromis holotaenia Trewavas
Haplochromis incola Trewavas
Haplochromis inornatus (Boulenger)
Haplochromis insignis Trewavas
Haplochromis intermedius (Guenther)
Haplochromis jacksoni Iles
Haplochromis johnstoni (Guenther)
Haplochromis kirkii (Guenther)
Haplochromis kiwinge Ahl
Haplochromis labidodon Trewavas
Haplochromis labifer Trewavas
Haplochromis labridens Trewavas
Haplochromis labrosus (Trewavas)
  (formerly Melanochromis
  labrosus)
Haplochromis lateristriga (Guenther)
Haplochromis lepturus Regan
Haplochromis leuciscus Regan
Haplochromis likomae Iles
Haplochromis linni Burgess & Axelrod
Haplochromis livingstoni (Guenther)
Haplochromis lobochilus Trewavas
Haplochromis longimanus Trewavas
Haplochromis macrostoma Regan

124

Haplochromis maculiceps Ahl
Haplochromis maculimanus Regan
Haplochromis marginatus Trewavas
Haplochromis melanonotus Regan
Haplochromis melanotaenia Regan
Haplochromis micrentodon Regan
Haplochromis microcephalus
  Trewavas
Haplochromis mloto Iles
Haplochromis modestus (Guenther)
Haplochromis mola Trewavas
Haplochromis mollis Trewavas
Haplochromis moorii (Boulenger)
Haplochromis nigritaeniatus
  Trewavas
Haplochromis nitidus Trewavas
Haplochromis nkatae Iles
Haplochromis nototaenia Boulenger
Haplochromis obtusus Trewavas
Haplochromis oculatus Trewavas
Haplochromis ornatus Regan
Haplochromis orthognathus Trewavas
Haplochromis ovatus Trewavas
Haplochromis pardalis Trewavas
Haplochromis phenochilus Trewavas
Haplochromis pholidophorus
  Trewavas
Haplochromis pictus Trewavas
Haplochromis placodon Regan
Haplochromis plagiotaenia Regan
Haplochromis pleurospilus Trewavas
Haplochromis pleurostigma Trewavas
Haplochromis pleurostigmoides Iles
Haplochromis pleurotaenia Boulenger
Haplochromis polyodon Trewavas
Haplochromis polystigma Regan
Haplochromis prostoma Trewavas
Haplochromis purpurans Trewavas
Haplochromis quadrimaculatus Regan
Haplochromis rhoadsii (Boulenger)
Haplochromis rostratus (Boulenger)
Haplochromis selenurus Regan
Haplochromis semipalatus Trewavas
Haplochromis serenus Trewavas
Haplochromis similis Regan
Haplochromis speciosus Trewavas
Haplochromis spectabilis Trewavas
Haplochromis sphaerodon Regan
Haplochromis spilonotus Trewavas
Haplochromis spilopterus Trewavas
Haplochromis spilorhynchus Regan
Haplochromis spilostichus Trewavas
Haplochromis stonemani Burgess &
  Axelrod
Haplochromis strigatus Regan
Haplochromis subocularis (Guenther)

Haplochromis taeniolatus Trewavas
Haplochromis tetraspilus Trewavas
Haplochromis tetrastigma Guenther)
Haplochromis triaenodon Trewavas
Haplochromis trimaculatus Iles
Haplochromis urotaenia Regan
Haplochromis venustus Boulenger
Haplochromis virgatus Trewavas
Haplochromis virginalis Iles
Haplochromis woodi Regan
Hemitilapia oxyrhynchus
  (Boulenger)
Iodotropheus sprengerae Oliver &
  Loiselle
Labeotropheus fuelleborni Ahl
Labeotropheus trewavasae Fryer
Labidochromis caeruleus Fryer
Labidochromis fryeri Oliver
Labidochromis mathothoi Burgess &
  Axelrod
Labidochromis textilus Oliver
Labidochromis vellicans Trewavas
Lethrinops alba Regan
Lethrinops alta Trewavas
Lethrinops argentea Ahl
Lethrinops aurita (Regan)
Lethrinops brevis (Boulenger)
Lethrinops christyi Trewavas
Lethrinops cyrtonotus Trewavas
Lethrinops furcicauda Trewavas
Lethrinops furcifer Trewavas
Lethrinops gossei Burgess & Axelrod
Lethrinops intermedia Trewavas
Lethrinops laticeps Trewavas
Lethrinops leptodon Regan
Lethrinops lethrinus (Guenther)
Lethrinops liturus Trewavas
Lethrinops longimanus Trewavas
Lethrinops longipinnis Eccles & Lewis
Lethrinops lunaris Trewavas
Lethrinops macracanthus Trewavas
Lethrinops macrochir Eccles & Lewis
Lethrinops macrophthalmus
  (Boulenger)
Lethrinops micrentodon Eccles &
  Lewis
Lethrinops microdon Eccles & Lewis
Lethrinops microstoma Trewavas
Lethrinops oculata Trewavas
Lethrinops parvidens Trewavas
Lethrinops polli Burgess & Axelrod
Lethrinops praeorbitalis (Regan)
Lethrinops stridei Eccles & Lewis
Lethrinops trilineata Trewavas
Lethrinops variabilis Trewavas
Lichnochromis acuticeps Trewavas

Haplochromis callipterus approaching full spawning coloration. The blue tones will intensify more yet. Photo by Dr. Warren E. Burgess.

Haplochromis callipterus in its normal coloration. The dark pattern on its sides and back is becoming visible. Photo by Dr. Warren E. Burgess.

Haplochromis placodon, female above, male below. This species is fairly common on sandy bottoms and in *Vallisneria* beds throughout the lake. It is a mollusc-crusher with heavy pharyngeal dentition. Photos by Dr. Herbert R. Axelrod.

127

Melanochromis auratus (Boulenger)
(formerly *Pseudotropheus auratus)*
Melanochromis exasperatus Burgess
Melanochromis johanni (Eccles)
(formerly *Pseudotropheus johanni)*
Melanochromis melanopterus
Trewavas
Melanochromis parallelus Burgess &
Axelrod
Melanochromis perspicax Trewavas
Melanochromis simulans Eccles
Melanochromis vermivorus
Trewavas
Petrotilapia tridentiger Trewavas
Pseudocrenilabrus philander (Weber)
Pseudotropheus aurora Burgess
Pseudotropheus brevis (Trewavas)
Pseudotropheus elegans Trewavas
Pseudotropheus elongatus Fryer
Pseudotropheus fainzilberi Staeck
Pseudotropheus fuscoides Fryer
Pseudotropheus fuscus Trewavas
Pseudotropheus lanisticola Burgess
Pseudotropheus livingstonii
(Boulenger)
Pseudotropheus lombardoi Burgess
Pseudotropheus lucerna Trewavas
Pseudotropheus macrophthalmus Ahl
Pseudotropheus microstoma Trewavas
Pseudotropheus minutus Fryer
Pseudotropheus novemfasciatus Regan

Pseudotropheus tropheops tropheops
Regan
Pseudotropheus tropheops gracilior
Trewavas
Pseudotropheus tropheops romandi
Columbe
Pseudotropheus tursiops Burgess
Pseudotropheus williamsi (Guenther)
Pseudotropheus zebra (Boulenger)
Rhamphochromis brevis Trewavas
Rhamphochromis esox (Boulenger)
Rhamphochromis ferox Regan
Rhamphochromis leptosoma Regan
Rhamphochromis longiceps (Guenther)
Rhamphochromis lucius Ahl
Rhamphochromis macrophthalmus
Regan
Rhamphochromis woodi Regan
Sarotherodon karongae (Trewavas)
Sarotherodon lidole (Trewavas)
Sarotherodon saka (Lowe)
Sarotherodon shirana (Boulenger)
Sarotherodon squamipinnis (Guenther)
Serranochromis robustus (Guenther)
Tilapia rendalli (Boulenger)
Tilapia sparrmani A. Smith
Trematocranus auditor Trewavas
Trematocranus brevirostris Trewavas
Trematocranus microstoma Trewavas
Trematocranus peterdaviesi Burgess
& Axelrod

There are many other "names" for Malawi cichlids in the literature, but for one reason or another they cannot be considered valid. Many are trade names, that is names coined by exporters or importers for use in identification but not validly described. In some cases there is more than one trade name applied to a fish, and commonly both an American and a European name exist.This trend has changed for the better with more descriptive names like "sulphur head", "sunset", "tangerine tiger", "electric blue", "red empress", "emerald green utaka", etc., which are descriptive, serve the purpose of identifying the fish, and do not cause problems for the scientific community.

A number of species have been inadequately described and the type specimens are such that identification at this time is uncertain. These species are under investigation and hopefully clarification of their status will be forthcoming. These are: *Melanochromis chipokae, M. interruptus, M. loriae, Microchromis zebroides, Pseudotropheus modestus, P. socolofi,* and *Trematocranus jacobfreibergi.* Two species proposed do not conform to the requirements adopted by the International Commission on Zoological Nomenclature for valid names and are here considered *nomina nuda.* These are *Labidochromis freibergi* and *L. joanjohnsonae.*

Haplochromis species males are even more difficult to identify when the breeding or dominant colors obscure the pattern. The bottom fish may be *Haplochromis incola*, but without examination of the pharyngeal teeth the identity must remain uncertain. Photos by Aaron Norman.

A non-dominant male of the *Haplochromis modestus* complex, probably the same species as shown opposite. Photo by Dr. Warren E. Burgess.

Breeding or dominant male *Haplochromis* sp. One must see this fish "in person" to appreciate its beauty. Photo by G. Marcuse.

Haplochromis sp. of the *modestus* complex. The dark pattern of this male obscures the spotted pattern that aids in its identification. Photo by Dr. Warren E. Burgess.

Haplochromis sp., possibly *H. mollis* or other diagonally striped *Haplochromis*. This is a male in dominant or breeding colors. Photo by Dr. Warren E. Burgess.

Haplochromis sp. in breeding colors. Most of the Lake Malawi cichlids have a great deal of blue color in their breeding or dominant phases. Photo by Hans Mayland.

Haplochromis atritaeniatus is one of the predatory species and occurs in the southern end of Lake Malawi and Upper Shire River. Photo by Dr. Herbert R. Axelrod.

A young male *Haplochromis electra* starting to gain his bright colors. Photo by Dr. Herbert R. Axelrod.

Haplochromis electra, the male (background) in his full breeding colors trying to coax the female (foreground) into spawning some more. Photo by T. Berardo.

Haplochromis auromarginatus is a shallow-water species that feeds off the bottom epiphytes. Photo by Dr. Herbert R. Axelrod.

Haplochromis longimanus was first collected by Christy from the southern end of Lake Malawi. Photo by Dr. Herbert R. Axelrod.

Haplochromis stonemani is a pretty little fish with a large eye. Photo by Dr. Herbert R. Axelrod.

Haplochromis decorus is one of the spotted species of *Haplochromis*. It was first discovered by Christy and described by Trewavas. Photo by Dr. Herbert R. Axelrod.

This species has been going under the name *Haplochromis quadrimaculatus*. Until specimens are obtained the identity will have to remain tentative. Photo by G. Meola, African Fish Imports.

Haplochromis sphaerodon attaining his dominant coloration. Photo by Dr. Herbert R. Axelrod.

Haplochromis species are difficult to identify. The pattern is similar in the two fishes on this page but not exactly alike (the bar is lower in the bottom fish). The dorsal fin spine count and the eye size are also different. Photos by Dr. Warren E. Burgess.

Haplochromis sp. is related to *H. incola*, but has a differently shaped head. Photo by Dr. Warren E. Burgess.

Haplochromis incola has enlarged, plate-like pharyngeal teeth, which indicates it is probably a mollusc-eater. Photo by Aaron Norman.

Haplochromis nototaenia (male above and female below) is one of the larger predatory cichlids of Lake Malawi, reaching a length of up to 40 cm. It lives on sand bottoms in sheltered areas. Photos by Dr. Herbert R. Axelrod.

Haplochromis ornatus is well named as you can see by this photo of a breeding male. It is reported to develop small lobes on the lips, but not as prominent as those of *H. euchilus*. Photo by Aaron Norman.

Haplochromis ornatus subdominant male with the lateral stripe showing. This species is a rock-frequenting species but technically not a true mbuna. Photo by Aaron Norman.

Haplochromis subocularis (probably a male). The eye has been discolored by the preservative (formalin) in which the fish was placed prematurely. Photo by Dr. Herbert R. Axelrod.

Haplochromis subocularis (?) female with the spotted band running diagonally from nape to caudal peduncle. A sand-dwelling species. Photo by Dr. Herbert R. Axelrod.

This specimen of *Haplochromis spilostichus* was trawled from 10-16 fathoms off Chembe village, Malawi. Photo by Michael K. Oliver.

Haplochromis ericotaenia also has the diagonal rows of spots from nape to caudal peduncle. Photo by Aaron Norman.

Haplochromis heterotaenia is a large predatory species found at all depths (to the limit of oxygen). It is more common in the northern sector of the lake. Photo by Michael K. Oliver.

Haplochromis epichorialis was first described by Trewavas from two specimens captured at Deep Bay by Christy. Photo by Michael K. Oliver.

Haplochromis anaphyrmus, full body photo
Photos by Dr. Herbert R. Axelrod.

Haplochromis obtusus male   showing diagonal stripe. Photo
by Dr. Herbert R. Axelrod.

Haplochromis hennydaviesi is an attractive species discovered on one of the early Axelrod collecting trips to Lake Malawi. Photo by Dr. Herbert R. Axelrod.

Haplochromis spilonotus has a horizontal band crossed by vertical bars. Photo by Dr. Herbert R. Axelrod.

A male *Haplochromis fenestratus* in full breeding dress. This is another rock-frequenting species of *Haplochromis*. It feeds on the organisms that inhabit the rock surfaces. Photo by Dr. Warren E. Burgess.

A young *Haplochromis fenestratus* in its fright pattern. Photo by Dr. Robert J. Goldstein.

Haplochromis strigatus from Lake Malawi. This is a sandy bottom or *Vallisneria* bed species. Photo by G. Meola, African Fish Imports.

Haplochromis strigatus is a predator on small fishes and insects but also feeds on some vegetable matter. Photo by Michael K. Oliver.

Haplochromis annectens has become one of the aquarium favorites, and locally raised fishes are now becoming available. Photo by Dr. Herbert R. Axelrod.

The breeding male *Haplochromis annectens* again has a predominantly blue color. Photo by Dr. Herbert R. Axelrod.

Haplochromis nigritaeniatus has a small mouth and a dark longitudinal stripe. Photo by Dr. Herbert R. Axelrod.

Haplochromis kiwinge also has a lateral stripe but a big mouth which indicates it is a predator. Its breeding colors are a beautiful blue with a golden spot on each scale. Photo by Dr. Herbert R. Axelrod.

Haplochromis similis male in breeding coloration. This is a fairly common species found not only in the lake proper but also in the estuaries of rivers emptying into the lake. It is a strict vegetarian feeding on submerged vegetation such as *Vallisneria*. The lower photo shows a close-up of the beautiful patterns of this male on the body and the vertical fins. Photos by Dr. Herbert R. Axelrod.

Haplochromis cf. *labridens,* probably a female, is another of the species characterized by a horizontal stripe. Photo by Dr. Herbert R. Axelrod.

Haplochromis similis male with a non-dominant color pattern. Photo by Dr. Herbert R. Axelrod.

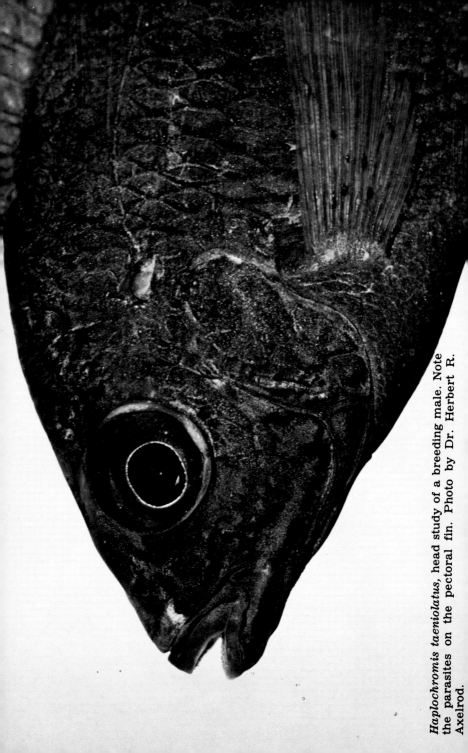

Haplochromis taeniolatus, head study of a breeding male. Note the parasites on the pectoral fin. Photo by Dr. Herbert R. Axelrod.

Haplochromis taeniolatus male (above) with a somewhat faded breeding pattern; female (below) showing longitudinal stripes. A pattern of cross bars is sometimes more evident than these longitudinal stripes. Photos by Dr. Herbert R. Axelrod.

Haplochromis labridens was also discovered in the southern parts of the lake. It is reportedly closely related to *Haplochromis kirki* but normally has one more dorsal fin spine (XVI) than that species (XV). Photo by Dr. Herbert R. Axelrod.

Haplochromis lepturus is a large predatory species over sandy bottoms near beaches. Photo by Dr. Herbert R. Axelrod.

A young *Haplochromis rostratus*. The pattern of spots is characteristic, as is an extended snout which develops as the fish grows. Photo by Dr. Herbert R. Axelrod.

Haplochromis woodi is a fairly large species (to 230mm) that is found in all parts of Lake Malawi. It has the elongated large-mouthed predatory look. Photo by Michael K. Oliver.

Haplochromis sp. (probably also *H. woodi*). The eye is relatively large in the younger fish shown here. Photo by Dr. Warren E. Burgess.

A fully grown *Haplochromis woodi* of an unusual color for a Lake Malawi cichlid. Photo by Dr. Herbert R. Axelrod.

Haplochromis modestus is another of the larger predatory cichlids of Lake Malawi. It is closely related to *H. woodi*. The upper photo shows a male in breeding coloration (photo by Dr. Warren E. Burgess), the lower photo shows the normal color (photo by Michael K. Oliver). Note the parasites (dark spots) on the lower fish.

157

Haplochromis pholidophorus is a very poorly known species. Until fairly recently it was known from a single specimen. Photo by Dr. Herbert R. Axelrod.

Haplochromis macrostoma (adult) is a large predatory cichlid from both ends of the lake although it is more common in the southern part. Photo by Dr. Herbert R. Axelrod.

A younger *Haplochromis macrostoma*. Photo by Dr. Herbert R. Axelrod.

A juvenile *Haplochromis fuscotaeniatus* has the color pattern of the adult females. Photo by Dr. Warren E. Burgess.

An adult male *Haplochromis fuscotaeniatus* in breeding colors. Photo by G. Meola, African Fish Imports.

Haplochromis polystigma male with faded colors. The pattern resembles *H. fuscotaeniata,* but note the spotted pectoral fins. Photo by Glen S. Axelrod.

The reaction of a male *Haplochromis polystigma* to its reflection is one of aggression. Photo by Glen S. Axelrod.

A male *Haplochromis polystigma* in breeding dress (above) and in its usual color pattern (below). It is interesting to note that the egg spots on the anal fin are only visible when the fish is not in breeding color. Photos by Dr. Herbert R. Axelrod.

A female *Haplochromis polystigma* in an aquarium. It is a rock-frequenting species. Photo by H. Hansen, Aquarium Berlin.

A male *Haplochromis polystigma* with faded breeding colors. Note the orange band along the edge of the anal fin. Photo by Dr. Herbert R. Axelrod.

Haplochromis livingstonii, head study. Photo by Dr. Herbert R. Axelrod.

Haplochromis livingstonii is a large predatory species that frequents beds of *Vallisneria* where it can hide from its prey. Photo by Dr. Herbert R. Axelrod.

Haplochromis livingstonii may "stalk" its prey by lying on its side on the bottom pretending to be dead until a curious fish gets too close. Photo by G. Marcuse.

A close-up of the characteristic head of *Haplochromis linni.* Photo by Dr. Herbert R. Axelrod.

A comparison photo showing two of the speckled species, *Haplochromis linni* (above) and *H. polystigma* (below). Photo by Dr. Herbert R. Axelrod.

A close-up of the head and body pattern of *Haplochromis liv-ingstonii*. Note the yellowish pectoral fins. Photo by Michael K. Oliver.

Haplochromis livingstonii in a typical "yawning" pose. Photo by G. Meola, African Fish Imports.

A male *Haplochromis venustus* in full breeding color. Photo by Dr. Herbert R. Axelrod at the Berlin Aquarium.

Haplochromis venustus male showing some of the blotchy pattern. This species frequents sandy areas and *Vallisneria* beds. Photo by G. Marcuse.

Haplochromis venustus, female with blotch pattern (above) and male (below). Note the sulphur yellow nape. Photos by Dr. Herbert R. Axelrod.

These two fishes belong to the genus *Haplochromis,* and in particular to the group of plankton-feeding species called "utaka." Their specific identity is uncertain (possibly *H. nkatae?*) Photo by Dr. Herbert R. Axelrod.

Haplochromis jacksoni is another member of the "utaka." Like the others it occurs in schools. Photo by Dr. Herbert R. Axelrod.

Haplochromis boadzulu has been found at White Rock and Boadzulu Island in Lake Malawi. This member of the "utaka" received its specific name from that island. Photo by Dr. Herbert R. Axelrod.

Haplochromis chrysonotus (above and below) showing the characteristic lateral spots. The lower fish is a male showing some of the blue color that develops during spawning. It is a "utaka." Photos by Dr. Herbert R. Axelrod.

Haplochromis mloto, male above, female below. Note the white to yellowish stripe from snout tip to dorsal fin. This is another of the zooplankton feeders of Lake Malawi. Photos by Dr. Herbert R. Axelrod.

Haplochromis virginalis is a zooplankton-feeder and is often referred to as "pure utaka," i.e., without spots. Photo by Dr. Herbert R. Axelrod.

Haplochromis eucinostomus is found over sandy bottoms and is similar to *H. mloto.* Both these "utaka" are called mloto by the natives although this species is distinguished as mloto mchenga by them. Photo by Dr. Herbert R. Axelrod.

Haplochromis argyrosoma, although similar in appearance to the utaka, has not been placed in that group by Iles, who reviewed the group. Photo by Dr. Herbert R. Axelrod.

Haplochromis johnstoni, one of the distinctly vertically barred *Haplochromis* species from Lake Malawi. Photo by G. Meola, African Fish Imports.

This laterally flattened species is *Haplochromis compressiceps*. It bears the name Malawian eye-biter although this may be largely exaggerated or even a misnomer. Photo by Dr. Herbert R. Axelrod.

A *Haplochromis compressiceps* male starting to develop its breeding coloration. Photo by Ken Lucas, Steinhart Aquarium.

The highly developed lips of *Haplochromis labrosus* may be sensory in nature and used for the detection of food. Photo above of a solid colored individual by Aaron Norman; photo below of the barred phase by Dr. Herbert R. Axelrod at the Berlin Aquarium.

This large-lipped *Haplochromis* sp. (probably a variety of *H. euchilus)* has affectionately been called the "Super VC-10" (after the jet plane) by Peter Davies. Juvenile above and adult male below. Photos by Dr. Warren E. Burgess.

Haplochromis euchilus with the normal color pattern of horizontal stripes. The adult has the enlarged lips. Photo by Dr. Herbert R. Axelrod.

A younger *H. euchilus,* probably a female, with the lips only slightly developed. Photo by H. Hansen, Aquarium Berlin.

Pseudotropheus elegans, female above, details of unpaired fins below. Photos by Dr. Herbert R. Axelrod.

The juvenile *Pseudotropheus elegans* shown here is very similar to the adults seen on these pages. Photo by Dr. Warren E. Burgess.

A male *Pseudotropheus elegans* approaching breeding color. Photo by Dr. Herbert R. Axelrod.

Haplochromis moorii is a sand-dweller. It was seen to follow a sand-digger (like *H. rostratus*) around snatching up bits of food that were uncovered. Photo by G. Marcuse.

The great development of the frontal hump in *Haplochromis moorii* is shown in this close-up of the head. Photo by Michael K. Oliver.

The young *Haplochromis moorii*, as seen here, is more normal in appearance although the first indications of the hump can be seen. Photo by Dr. Herbert R. Axelrod.

Lethrinops furcicauda, probably a female. This is one of the sand-feeders. The high position of the eye makes this possible without damage to the eye. Photo by Dr. Herbert R. Axelrod.

A close-up of the anal fin of *Lethrinops furcicauda.* Photo by Dr. Herbert R. Axelrod.

An as yet unidentified species of *Lethrinops*. It is possibly new to science. Photo by Dr. Warren E. Burgess.

Lethrinops furcicauda male from Lake Malawi. The food consists of crustaceans, molluscs, insect larvae, and other items associated with the sand bottom. Photo by Dr. Herbert R. Axelrod.

Lethrinops parvidens is quite common in the southeastern part of the lake. Photo by Dr. Herbert R. Axelrod.

Lethrinops aurita in its non-breeding colors is plain silvery, an adaptation for existence over a sand bottom. Photo by Dr. Herbert R. Axelrod.

Lethrinops sp., probably near *L. lethrinus*. Only a few species of *Lethrinus* are shipped from Lake Malawi. Photo by Aaron Norman.

Lethrinops lethrinus is more common in the southern end of Lake Malawi where it was first collected by Johnston. Photo by Dr. Herbert R. Axelrod.

Lethrinops christyi is quite a spectacular fish. Perhaps when it can be found in sufficient quantities it will be included in the Malawi shipments. Photo of a male (above) and a closeup of the iridescent colors of the head (below). Photos by Dr. Herbert R. Axelrod.

Lethrinops gossei is similar in appearance to *L. christyi* but distinguishable at once by the distance from the eye to the mouth. Photos of a male (above) and a female (below) by Dr. Herbert R. Axelrod.

Lethrinops polli is one of the more colorful of the species of *Lethrinops*. Photo by Dr. Herbert R. Axelrod.

Trematocranus peterdaviesi was described by the authors back in 1973. Photo by Dr. Herbert R. Axelrod.

Above and below: *Trematocranus* sp. This fish has been sold under the trade name "Red-Top *Aristochromis*" even though there is no resemblance at all between *Trematocranus* and the true *Aristochromis*. Photos by G. Meola, African Fish Imports.

Trematocranus jacobfreibergi has become one of the most poular of the Lake Malawi cichlids. The two photos of the males (above and below) show why this is so. Photo above by Dr. Herbert R. Axelrod; photo below by G. Meola, African Fish Imports.

A young male *Trematocranus jacobfreibergi* just starting to develop some of the coloring of the adult. Photo by Dr. Warren E. Burgess.

Female *Trematocranus jacobfreibergi* are banded and not very colorful. Photo by Dr. Warren E. Burgess.

A breeding male *Aulonocara macrochir* trawled from deep water in Lake Malawi. Photo by Michael K. Oliver.

The "night *Aulonocara*" (probably *Aulonocara nyassae* var.), reaching a length of only 3-4", is the smallest of the peacock blue morphs. Photo by G. Meola, African Fish Imports.

The "yellow *Aristochromis*" is not a species of *Aristochromis* but a morph of *Aulonocara nyassae.* A better name would be yellow peacock or yellow *Aulonocara.* Photo above by G. Meola, African Fish Imports; photo below by Hans Mayland.

The peacock blue cichlid, *Aulonocara nyassae*. This is another of the aquarium favorites because of the bright colors. It has been bred in sufficient quantities to make it readily available. Photo by Aaron Norman.

A peacock blue cichlid, *Aulonocara nyassae,* of the solid blue variety in contrast to the orange-shouldered variety shown above. Photo by G. Meola, African Fish Imports.

A young male *Aulonocara nyassae* (note the large eye). The species of *Aulonocara* and *Trematocranus* all have enlarged head pores which can be seen with the naked eye in an aquarium. Photo by Dr. Warren E. Burgess.

The females and young male *Aulonocara nyassae* do not exhibit the brilliant colors of the dominant or breeding males, but retain this barred pattern and are brownish in color. Photo by Dr. Warren E. Burgess.

A male *Melanochromis melanopterus* photographed in the lobby
aquarium of the Monkey Bay Hotel on Monkey Bay, Malawi.
Photo by Dr. Herbert R. Axelrod.

A second male *Melanochromis melanopterus* with more yellow in
its fins. Note the reduced number of anal fin spots. Photo by Dr.
Herbert R. Axelrod.

Melanochromis sp. The genus *Melanochromis* is one of the least understood of the mbuna genera. This might be *M. simulans*. Photo by Dr. Warren E. Burgess.

Melanochromis melanopterus with an almost solid black body. The yellow on the edges of the vertical fins makes a pleasing contrast. Photo by Dr. Warren E. Burgess.

Female *Melanochromis* "chipokae" (above) and male of the same species (below) showing the typical *Melanochromis* patterns, i.e. dark stripes on a light background in the female and light stripes on a dark background in the male. Photos by G. Meola, African Fish Imports.

A male (dark fish) and a female *Melanochromis parallelus*. This species has been known in the trade as the black-and-white auratus. Photo by Dr. Herbert R. Axelrod.

Melanochromis parallelus male. The dorsal fin is black and the lateral stripes bluish in this species. Photo by Dr. Warren E. Burgess.

The Likoma Island morph of a male *Melanochromis johanni* (above) differs in pattern somewhat from the individual from another part of the lake shown opposite. The female (below) does not seem to differ very much (if at all). Photos by Dr. Warren E. Burgess.

Melanochromis johanni male (above) and female (below). The male has a solid upper stripe but a broken lower stripe. The females and young males are a solid yellow. Photos by Uwe Werner.

These two individuals were captured by Peter and Henny Davies and described as a new species, *Pseudotropheus johanni*, by Eccles. The species was later shifted to the genus *Melanochromis*. The female (above) was designated as holotype, and the male (below) a paratype. Photos by Dr. Herbert R. Axelrod.

A *Melanochromis johanni* male lacking the lateral white stripes. Note the few anal fin spots placed well back on the fin. Photo by H.-J Richter.

An unusual group of *Melanochromis johanni* males. Usually a single male will dominate the tank and other males take on a subdued pattern. Photo by Dr. Herbert R. Axelrod at the Berlin Aquarium.

The fishes above and below and on the opposite page appear to be another morph of *Melanochromis johanni* and have been called *M.* "ornatus" in the trade. These males are very much like those on the previous pages except for the spotted caudal fin. Photos by Dr. Herbert R. Axelrod.

A female (above) and male (below) *Melanochromis johanni* from Likoma Island, Lake Malawi. Females occasionally assume a weak male coloration to avoid attention by males. Photo by Dr. Herbert R. Axelrod.

Two female *Melanochromis johanni* with more normal colors. This Likoma Island form has a dark bar in the dorsal fin as well as the spots in the caudal fin. Photo by Dr. Herbert R. Axelrod.

A male *Melanochromis vermivorus* in dominant or breeding color. This color and pattern is very much like that of other species of *Melanochromis*. Photo by Hilmar Hansen, Aquarium Berlin.

A male *Melanochromis vermivorus* that has been caught by the camera in an intermediate phase. It is difficult to believe the color changes that the males go through unless you see it yourself or have a photo such as this one. Photo by Hilmar Hansen, Aquarium Berlin.

The typical coloration of a female *Melanochromis vermivorus*. Young males also have this pattern but change when they reach sexual maturity. Photo by G. Marcuse.

A male *Melanochromis vermivorus* is recognizable in part by the white dorsal fin. This species is one of the more aggressive species of lake cichlid. Photo by Dr. Warren E. Burgess.

Melanochromis simulans male. This species has a longer snout than *M. vermivorus,* which helps distinguish it from that similar looking species. Photo by G. Marcuse.

Melanochromis perspicax dominant or breeding male (above) and female (below). It has been said that this fish is not the real *M. perspicax* but a new species, so the identification remains tentative until further investigations are completed. Photos by Dr. Warren E. Burgess.

A dominant or breeding male *Melanochromis auratus* (above) and a normal female (below). This species was one of the first popular African lake cichlids of the *Melanochromis* type. Photos by Dr. Herbert R. Axelrod.

It is very easy to recognize the male *Melanochromis auratus* by his color pattern (above), but young males and females (below) look alike and cannot easily be distinguished. "Females" often decide to change colors and become males. Photos by Dr. Warren E. Burgess.

Even in *Melanochromis auratus* one can find different color morphs. The lateral stripe of the male might be tan to white and the dorsal fin white to yellowish; the female might be a bright yellow or an orange-yellow, all depending upon the collection locality. Photos (male above, female below) by Dr. Herbert R. Axelrod.

Melanochromis simulans has a very similar color pattern to that of *M. auratus*. The more pointed snout in *M. simulans* makes them distinguishable. The bicolored pattern of the caudal fin in *M. auratus* helps separate the females. Photos by Dr. Herbert R. Axelrod.

Melanochromis exasperatus is another of those species that do not fit well into any particular genus. Until further delimitation of generic boundaries is possible it will remain where it was described. Photos of a male (above) and female (below) by Dr. Herbert R. Axelrod.

As in other *Melanochromis* species the young male and female *Melanochromis exasperatus* (above) are identical; the males (below) take on a different pattern and coloration. This is a very popular aquarium species. Photos by Dr. Herbert R. Axelrod.

These fishes (above and below) have been difficult, if not impossible, to identify without specimens. The upper fish may be *Pseudotropheus brevis* (changed from *Melanochromis brevis*); the bottom one is called *Pseudotropheus* "teyoi" but may not even be a species of *Pseudotropheus*. Photos by Aaron Norman.

The shape of the snout distinguishes *Pseudotropheus tursiops* from other species of *Pseudotropheus*. Photo by Dr. Herbert R. Axelrod.

A species of *Pseudotropheus* commonly called *Haplochromis formosus*. It acts and breeds like a typical *Pseudotropheus* species. Photo by Aaron Norman.

Photo tank specimens of *Pseudotropheus elongatus,* male (above) and female (below), collected by Peter Davies at Lake Malawi. Photos by Dr. Herbert R. Axelrod.

Aquarium specimens of *Pseudotropheus elongatus*. The lower, more brightly colored fish is the male, the upper, paler fish is the female. Photos by Dr. D. Terver, Nancy Aquarium, France.

Pseudotropheus sp. of the *P. elongatus* complex. This fish has been sold under various names in the trade such as *P.* "bentoni". Photo by Dr. Warren E. Burgess.

A male *Pseudotropheus* sp., possibly the same species as the one above. Note the very long pelvic fins. Photo by Dr. Herbert R. Axelrod.

Pseudotropheus sp. The large yellow "egg" spots on the posterior part of the anal fin are matched by similar yellow spotting in the dorsal fin. Photo by Dr. Herbert R. Axelrod.

Pseudotropheus elongatus male from Lake Malawi. The *P. elongatus* complex is still confused and needs work before the species are all straightened out. Photo by Karl Knaack.

A variety of *Pseudotropheus elongatus* with the slender body but with a spotted caudal fin. Photo by Dr. Warren E. Burgess.

A possible new species of *Pseudotropheus* from Lake Malawi. As work progresses in the study of the Lake Malawi cichlids, new species will be described. Photo by Dr. Herbert R. Axelrod.

An unidentified species of *Pseudotropheus* in the *P. elongatus* or *P. tropheops* complexes. The normal blue form (above) and a yellow form (below) are available according to dealers. Photo above by Dr. Warren E. Burgess, photo below by G. Meola, African Fish Imports.

Pseudotropheus sp., one of the many unidentifiable mbunas from Lake Malawi. Photo by Glen S. Axelrod.

Pseudotropheus livingstonii has often been confused with *P. lanisticola* but is easily recognized by the lack of a bright yellow bar on the anal fin. Photo by G. Meola, African Fish Imports.

This species has been going under the name *Pseudotropheus* "chameleo" in the trade and is possibly a new species. Photo by S. Kochetov.

A male *Pseudotropheus* "chameleo." There are reports that a second "chameleo" with different banding is also being imported from Lake Malawi. Photo by Aaron Norman.

This is a photograph of the specimen used as the holotype of the shell-dwelling species *Pseudotropheus lanisticola*. Photo by Dr. Warren E. Burgess.

The caudal fin pattern of *Pseudotropheus lanisticola* is very distinctive for an mbuna. Photo by Dr. Herbert R. Axelrod.

Pseudotropheus lanisticola lives in an area where there are few rocks. It hides in empty shells of the genus *Lanistes*. It is quite surprising how a relatively large individual can squeeze itself into a relatively small shell. Only the tail of the fish can be seen in the shell below. Photo above by G. Meola, African Fish Imports; photo below by Dr. Warren E. Burgess.

Pseudotropheus sp. This light blue mbuna goes under a number of names *(Pseudotropheus* "eduardi," *P.* "pindani," *P. socolofi, P.* "newsi," etc.), none of which may be valid. Photo by A. Kochetov.

The caudal fin has each ray accented with black and there is a black band in the anal fin. In some individuals there is a black band in the dorsal fin. Photo by Dr. Herbert R. Axelrod.

This fish is similar in pattern to *P.* ''eduardi'' and *Labidochromis fryeri* but with different tooth and color characteristics. Photo by Dr. Herbert R. Axelrod.

There are no obvious differences between the sexes in *Pseudotropheus* ''eduardi.'' Photo by G. Meola, African Fish Imports.

An unknown species of mbuna, probably a *Pseudotropheus*, called *P.* "jacksoni" in the trade. Photo by G. Meola, African Fish Imports.

Pseudotropheus minutus is shaped something like a *Labidochromis* but does not have the tooth structure of that genus as currently defined. Photo by Dr. Herbert R. Axelrod.

Pseudotropheus minutus (above and below). The above individual has more bars on the body and the one below has an indication of a dorsal fin stripe, but both appear to be the same species. Photo above by Dr. Warren E. Burgess; photo below by G. Meola, African Fish Imports.

Pseudotropheus aurora is quite colorful, with contrasting blue and yellow areas. This species was erroneously called *P. lucerna* when it was first imported. Photo by Aaron Norman.

Pseudotropheus sp. No definite identification of this fish has as yet been possible. Photo by Dr. Herbert R. Axelrod.

Pseudotropheus aurora has a very large eye which helps distinguish it from other *Pseudotropheus* species. Note also the yellow dorsal fin and single yellow anal fin spot. Photo by Dr. Herbert R. Axelrod.

A pair of *Pseudotropheus lombardoi,* male (above) and female (below). Photo by G. Meola, African Fish Imports.

An intermediate phase of a male changing from the bluish color to the yellow. Photo by Glen S. Axelrod.

Male *Pseudotropheus lombardoi* (also called "kennyi" or "lilancinius" in the trade) with its golden yellow color, a reversal of the normal *Pseudotropheus* pattern. Photo by Ed Isaacs, Pet Gallery.

The female *Pseudotropheus lombardoi* is blue, also a reverse of the normal situation of male and female *Pseudotropheus* colors. Photo by Ed Isaacs, Pet Gallery.

A male *Pseudotropheus microstoma* from Lake Malawi. Photo by Dr. R. J. Goldstein.

The female *Pseudotropheus microstoma* is less colorful than the male. Photo by Dr. R. J. Goldstein.

Pseudotropheus microstoma, dark phase. The steeply descending snout marks this as a member of the *Pseudotropheus tropheops* complex. Photo by Dr. Warren E. Burgess.

This fish is a rather slender mbuna but undoubtedly a member of the *Pseudotropheus tropheops* complex. Photographed in Lake Malawi by Dr. Herbert R. Axelrod.

Males of many of the cichlids are often more colorful than the females. The upper fish appears to be the male and the lower one the female. This is probably *Pseudotropheus brevis* (formerly *Melanochromis brevis*). Photos by Dr. Warren E. Burgess.

This mottled female *Pseudotropheus tropheops* has a very colorful background of yellowish orange with blue edging to the scales. Photo by Dr. Herbert R. Axelrod.

Pseudotropheus tropheops female from Likoma Island, Lake Malawi. Photo by Dr. Warren E. Burgess.

A pair of *Pseudotropheus tropheops*, female above, male below. The *P. tropheops* complex is badly in need of a revision. Photos by Dr. Warren E. Burgess.

A *Pseudotropheus tropheops* variety from Likoma Island, Lake Malawi, male above and female below. The anal fin spots are usually more developed in males. Photos by Dr. Warren E. Burgess.

The two specimens of *Pseudotropheus tropheops* shown on this page are more slender-bodied than those on adjacent pages and have a rounded tooth band. These differences, among others, were used by J. Colombé as a basis for describing it as a new subspecies, *P. tropheops romandi*. Photos by Dr. Herbert R. Axelrod.

A breeding male *Pseudotropheus tropheops*. This blue morph is not common in the U.S. aquarium trade. Photo by Klaus Paysan.

A bright yellow morph of *Pseudotropheus tropheops*. There seems to be a large number of morphs of *P. tropheops* already known and more being discovered all the time. Photo by Dr. Herbert R. Axelrod.

One of the newer varieties of *Pseudotropheus tropheops*. Not only are new varieties being discovered in Lake Malawi, but cross-breeding by aquarists is producing additional domestic forms. Photo by Aaron Norman.

A yellow morph of *Pseudotropheus tropheops*. Note the indication of horizontal stripes, unusual in the genus *Pseudotropheus*. Photo by Andre Roth.

A yellow morph of *Pseudotropheus tropheops* with black markings in its fins. Photo by Dr. Herbert R. Axelrod.

Pseudotropheus tropheops, male above, female below, collected near Likoma Island in Lake Malawi. Photo by Dr. Herbert R. Axelrod.

This morph of *Pseudotropheus tropheops* has an orange shoulder blotch. Photo by G. Meola, African Fish Imports.

A male *Pseudotropheus tropheops*. This form with the orange shoulder is said to be the form used by Regan in his description of the species. Photo by Dr. Herbert R. Axelrod.

Pseudotropheus sp. (possibly *P. tropheops* or a member of the *P. tropheops* complex). This fish (above and below) is a male. Again the anal fin spots are way back in the fin. Photos by Andre Roth.

Pseudotropheus sp. (probably *P. williamsi*). *Pseudotropheus williamsi*, the type species of the genus *Pseudotropheus*, has these dark spots along its sides and back. Until specimens can be obtained the identification has to remain in question. Photos by Klaus Paysan.

In the B morph of *Pseudotropheus zebra* the fish range in color from sky blue to a mauve, both sexes being colored alike. It is the opinion of some people that some of the solid color "zebras" are a different species. Photos by Dr. Herbert R. Axelrod.

White morphs of *Pseudotropheus zebra* are very popular in the aquarium trade. White males (below), however, are not common and are difficult to spot, therefore commanding a much higher price than females (above). Photo above by G. Meola, African Fish Imports; photo below by Dr. Herbert R. Axelrod.

There are several morphs of *Pseudotropheus zebra* that are more or less orange in color. They may be called red zebras, tangerine zebras, orange zebras, etc. In some instances the two sexes are differently colored, only one sex ever being red. Photo by Dr. Herbert R. Axelrod.

The green morph of *Pseudotropheus zebra* is probably just an extension of the blue morph. Photo by G. Marcuse.

This morph is also called the green zebra but is different from the solid colored one on the opposite page. This male is a variation on the BB type. Photo by Aaron Norman.

An old but still impressive *Pseudotropheus zebra* male. Note the hump on the nape and the very elongate pelvic fins. Photo by Klaus Paysan.

Typical female *Pseudotropheus zebra* have a blotched pattern
and are referred to as OB females for orange blotch. The blotching may be strong (above) or very weak (below). Photos by Dr.
Herbert R. Axelrod.

An unusually colorful *Pseudotropheus zebra* female. This individual had 19 ripe eggs in her right ovary. Photo by Dr. Herbert R. Axelrod.

A brown zebra morph. This *Pseudotropheus zebra* male is heavily infested with parasites (dark spots). Photo by Dr. Warren E. Burgess.

A blotched zebra with the blotches small or broken up is referred to as a spotted zebra. Photo by G. Meola, African Fish Imports.

Pseudotropheus zebra mottled or blotched form but with peppering of dark spots, a variation in the OB female. Photo by Dr. Warren E. Burgess.

Two variations on the *Pseudotropheus zebra* OB. As can be seen, the blotching is highly variable. A male OB (above) is rare and costly. It is called a marmalade cat and usually has a bluish sheen on the body that the female OB's lack. Photos by G. Meola, African Fish Imports.

A normal BB *Pseudotropheus zebra*. The BB stands for black and blue, referring to the dominant pattern. Photo by H. Hansen, Aquarium Berlin.

A black-top *Pseudotropheus zebra*. Note the black "mask" on the face and the yellow branchiostegal membrane at the throat. Photo by Dr. Herbert R. Axelrod.

A red-top *Pseudotropheus zebra*. The dorsal fin is more orange than red but still makes for a colorful morph. Photo by Dr. Herbert R. Axelrod.

Pseudotropheus zebra morph with many bars on the sides. Note also the very elongate pelvic fins. Photo by S. Kochetov.

Cynotilapia afra looks to all intents and purposes like *Pseudotropheus zebra*, but smaller. However, in *Cynotilapia* the teeth are unicuspid while in *Pseudotropheus* they are bicuspid in the outer row. Photo by Dr. Herbert R. Axelrod.

A male (upper fish) and female (lower fish) *Cynotilapia afra* captured in Lake Malawi near Likoma Island. Photo by Dr. Herbert R. Axelrod.

A colorful *Cynotilapia* species from the Mozambique shores of Lake Malawi. Photo by Dr. Warren E. Burgess.

A *Cynotilapia afra* morph with an interesting dorsal fin pattern. Photo by G. Meola, African Fish Imports.

Cynotilapia afra morph commonly called "eduardi" or flagfin *Cynotilapia*. Photo by Dr. Herbert R. Axelrod.

A variation on the "eduardi" or flagfin morph of *Cynotilapia afra* with whitish instead of yellowish in the dorsal fin. Photo by Dr. Warren E. Burgess.

Cynotilapia axelrodi female has less intense coloration than the male. This female was nearly ripe. Photo by Dr. Warren E. Burgess.

Cynotilapia axelrodi male. The more intense colors and the pattern of the anal fin help distinguish the sexes in this species. Photo by Dr. Warren E. Burgess.

A very colorful individual of *Iodotropheus sprengerae*. This species has been bred in enough quantity to make domestic individuals readily available. Photo by Stanislav Frank.

The more normal coloration of *Iodotropheus sprengerae* from which it received the common name rusty cichlid. Photo by Jaroslav Elias.

Two new imports from Lake Malawi which have not been identified as yet. They are probably species of the genus
Gephyrochromis. Photos by Dr. Warren E. Burgess.

Two male *Labidochromis vellicans*. The relatively shallow body, quite visible in this photo, is one of the species' characteristics. Photo by Dr. Herbert R. Axelrod.

Labidochromis cf *vellicans*. *Labidochromis* species have elongated unicuspid teeth for picking insects and small crustaceans from the rock biocover. Photo by Dr. Warren E. Burgess.

Labidochromis mathothoi, male above, female below. The narrow head prevents species of *Labidochromis* from incubating large numbers of young. Photo by Dr. Herbert R. Axelrod.

Labidochromis sp. male. This is probably a new species or a variant of *Labidochromis mathothoi.* Its dull coloration will probably keep it from being imported in any great quantity. Photo by Dr. Warren E. Burgess.

A male (above) and female (below) *Labidochromis* sp. that has been dubbed "horseface Labidochromis" by hobbyists. Note the large spots in the anal fin of the male. Photos by Jim Seidewond.

A new *Labidochromis* from Mumbo Island. This pattern is seen in other mbuna but not in any other known *Labidochromis* species. Photo by Dr. Warren E. Burgess.

Labidochromis sp., probably what has been called *L. freibergi*, although the species name is probably not valid. Photo by Dr. Warren E. Burgess.

Labidochromis caeruleus consistently has dark bars in the dorsal and anal fins. The body color is normally silvery white with multicolor reflections. Photo by G. Meola, African Fish Imports.

Labidochromis caeruleus, breeding male. The bright blue color of the breeding male was the origin of the specific name. Photo by Dr. Warren E. Burgess.

Labidochromis fryeri from Mumbo Island. This species is difficult to photograph in an aquarium because of the reflective nature of the pearly white scales. Photo by H. Hansen, Aquarium Berlin.

Labidochromis fryeri male. The breeding male, like many other lake cichlids, exhibits much more blue than the female. Photo by Dr. Warren E. Burgess.

Labidochromis fryeri has a dark band in the anal fin but lacks a similar band in the dorsal. This helps distinguish it from *L. caeruleus*. Photo by Dr. Herbert R. Axelrod.

This photo is one of the original specimens of *Labidochromis fryeri*. The fish apparently came from Mumbo Island, Lake Malawi and was exhibited in an aquarium in the lobby of the Monkey Bay Hotel. Photo by Dr. Herbert R. Axelrod.

Labidochromis textilis has been involved in the confusion between *L. fryeri, L. joanjohnsonae,* and *Melanochromis exasperatus.* Photo by G. Meola, African Fish Imports.

The female *Labidochromis textilis* is distinguishable in an aquarium from *Melanochromis exasperatus* females by its more numerous and more even horizontal stripes. Photo by Dr. Bruce J. Turner.

The Malawian scale-eater, *Genyochromis mento*. Females have the typical OB or orange blotch pattern. Photo by A. Ivanoff.

The male Malawian scale-eater. The common name is well deserved; not many tankmates are safe from its attack. Photo by A. Ivanoff.

One of the more colorful morphs of *Labeotropheus fuelleborni,* the orange-sided fuelleborni. Photo by G. Meola, African Fish Imports.

Labeotropheus fuelleborni male. The snout may diminish in size and eventually disappear with aquarium inbreeding. Photo by Dr. D. Terver, Nancy Aquarium, France.

The black-top fuelleborni is another of the common morphs of *Labeotropheus fuelleborni* available to the hobbyist. Photo by Dr. Herbert R. Axelrod.

The contrasting orange dorsal fin (referred to as a red-top) and blue body of this male *Labeotropheus fuelleborni* make this a very popular variety. Photo by Dr. Herbert R. Axelrod.

A white morph of *Labeotropheus fuelleborni* with some blotching. Chances are this is a female, as male whites are hard to find. Photo by G. Meola, African Fish Imports.

A colorful *Labeotropheus fuelleborni* female OB. The teeth of *Labeotropheus* species are all tricuspid, an adaptation for scraping algae from rocks. Photo by Andre Roth.

A more heavily blotched female OB *Labeotropheus fuelleborni*. The blotching varies from almost non-existent to very heavy in OB morphs. Photo by Dr. Warren E. Burgess.

A variation on the OB called the gold blotch *Labeotropheus fuelleborni*. Note that spotting is present even on the pelvic fins of some of these morphs. Photo by G. Meola, African Fish Imports.

A marmalade cat *Labeotropheus fuelleborni,* but without the red top. These males can often be recognized by some bluish tints on the body. Photo by G. Meola, African Fish Imports.

One of the most prized morphs of the genus *Labeotropheus* is the red-top marmalade cat shown here. It is a male blotched form with an orange-red dorsal fin. Photo by G. Meola, African Fish Imports.

A male (above) and female (below) *Labeotropheus fuelleborni.*
When similarly patterned, the males usually show the more intense colors, especially when breeding. The deeper body helps
distinguish them from the other species of the genus,
Labeotropheus trewavasae. Photos by Dr. Herbert R. Axelrod.

An atypical morph of *Labeotropheus trewavasae*. Certainly not all of the different color varieties of these species have been discovered as yet. Photo by Dr. Herbert R. Axelrod.

Labeotropheus trewavasae. This is one of the more common combinations of male/female *L. trewavasae* sold today, the red-top male (upper fish) and the OB female (lower fish). Photos by Dr. D. Terver, Nancy Aquarium, France.

A darker version of the red-top in which the dorsal fin is closer to a true red. The vertical bars usually signify that the fish is frightened. Photo by Andre Roth.

A male red-top *Labeotropheus trewavasae*. The anal fin spots are also located in the back part of the fin in this genus. Photo courtesy of Wardley Products Co.

A male (darker fish at back) and female (yellow fish) *Labeotropheus trewavasae*. The golden trewavasae are quite popular among hobbyists. Photo by Dr. Herbert R. Axelrod.

A more intensely colored *Labeotropheus trewavasae* female. The intensity of the colors will vary considerably even in the same brood. Photo by Dr. Herbert R. Axelrod.

A mottled male *Labeotropheus trewavasae* (above). It is difficult to distinguish this marmalade cat from the normal OB female (below). An examination of the genital area is necessary to identify the sex. Photos by Dr. Herbert R. Axelrod.

Petrotilapia tridentiger is probably one of the largest of the mbunas. It reaches a length of about 20 cm in nature. Photo by Dr. Warren E. Burgess.

Petrotilapia tridentiger penetrates to rather deep water, almost reaching the limits of the dissolved oxygen. Photo by Dr. Herbert R. Axelrod.

A large male *Petrotilapia tridentiger*. The mouth is filled with small tricuspid teeth (hence the name "tridentiger") which are used for scraping algae off rocks. Photo by Dr. Herbert R. Axelrod.

Normal patterns of *Petrotilapia tridentiger*. Note that the mouth always looks as if it cannot completely close. Two males and a female (light colored) can be seen here. Photo by Dr. R. J. Goldstein.

This fish was collected in relatively deep water. It was named *Cleithrochromis bowleyi* by D. H. Eccles. Photo by Dr. Herbert R. Axelrod.

Tilapia sparrmani has been found in waters near Lake Malawi but as yet has not been found in the lake itself. It is omnivorous and is one of the few nesting species of cichlid in the lake. Photo by Dr. R. J. Goldstein.

Hemitilapia oxyrhynchus is a browser among *Vallisneria* beds, scraping algae and insects from the leaves. Photo by Dr. Herbert R. Axelrod at the Berlin Aquarium.

Hemitilapia oxyrhynchus may grow to 20 cm but is usually smaller. The triangular black spots are distinctive. Photo by Dr. Herbert R. Axelrod.

Chilotilapia rhoadesii male in normal everyday colors (above) and breeding colors (below). This species is a mollusc-crusher and as such has heavy teeth. Photos by Dr. Herbert R. Axelrod.

A healthy aquarium specimen of *Chilotilapia rhoadesii* with the typical barring of the female and young males. Photo by Aaron Norman.

A young male *Chilotilapia rhoadesii* which is starting to develop a dominant or breeding coloration especially visible about the head. Photo by G. Meola, African Fish Imports.

Species of the genus *Rhamphochromis* are predators, feeding on cichlids of the "utaka" group as well as *Engraulicypris sardella,* a cyprinid. Top: *Rhamphochromis* sp.; Center: *Rhamphochromis macrophthalmus.* Photos by Dr. Herbert R. Axelrod.

One of the large-mouthed predators, *Serranochromis robustus.* It is more characteristic of estuaries and sheltered lake waters but can be found off rocks in the open lake. It reaches a length of up to 50 cm. Photo by Dr. Herbert R. Axelrod.

Diplotaxodon argenteus has been captured at depths of 30-40 meters where it feeds on other fishes, mainly *Engraulicypris*. Photo by Dr. Herbert R. Axelrod.

Diplotaxodon ecclesi is similarly a predator in the lake, frequenting fairly deep waters. Photo by Dr. Herbert R. Axelrod.

Cyprichromis leptosoma (above and below). The difference in the color of the caudal fins has not been explained. Some people feel that there is more than one species involved here. Photos by Glen S. Axelrod.

Young *Cyprichromis brieni* (above and center). Recent study of fishes in this genus turned up this new species, increasing the total to four. Photo above by Dr. Warren E. Burgess; center photo by Pierre Brichard.

Cyprichromis brieni, like the others in this genus, is a schooling fish feeding mainly on the zooplankton (copepods, etc.). Photo by Pierre Brichard.

Cyprichromis microlepidotus is easily recognizable by the very small scales. The *Cyprichromis* are mouthbrooders and come into shallower water to brood and release the fry. Photo by Pierre Brichard.

Lestradea perspicax perspicax feeds on the microorganisms found in the mud and sand. It is a mouthbrooder. Photo by Pierre Brichard.

The ventral fins of *Cyathopharynx furcifer* are very elongate and are provided at their tips with a bright yellow spot which some people are comparing with the "egg" spots of the *Haplochromis* species. It is also a mouthbrooder. Photo by Glen S. Axelrod.

Aulonocranus dewindti occurs in shallow water over sand or rocky outcrops in sand. It is a mouthbrooder. Photo by Pierre Brichard.

Aulonocranus dewindti. This genus should not be confused with *Aulonocara* of Lake Malawi. Photo by Glen S. Axelrod.

Cardiopharynx schoutedeni is common along sandy beaches. It is a mouthbrooder with up to 60 eggs in a batch. Photo by Glen S. Axelrod.

Ophthalmochromis ventralis has been divided into two subspecies. This specimen from the southern part of the lake is *O. ventralis heterodontus.* Photo by Pierre Brichard.

Ophthalmochromis ventralis is a mouthbrooder which, like *Cyathopharynx,* has elongate ventral fins provided with yellow spatulate tips. Photo by Dr. Herbert R. Axelrod at the Berlin Aquarium.

A young *Cyphotilapia frontosa*. The nuchal hump has not yet developed. The development can be seen in the following photos. More young *Cyphotilapia frontosa* are becoming available as they are being bred domestically in larger and larger numbers. This is helpful since wild specimens are still quite expensive. They come from deep water and suffer from the bends, so the losses are high compared to the shallow-water species. Photo by H.-J. Richter. Adult (below) by Dr. Herbert R. Axelrod.

Cyphotilapia frontosa are mouthbrooders, but instead of brooding and releasing the fry in shallow water like most other deep-water mouthbrooders, they remain in deep water. Photo by H. Hansen, Aquarium Berlin.

An adult male *Cyphotilapia frontosa* with a well-developed hump. This species is often compared to *Haplochromis moorii* of Lake Malawi because of the similar nuchal hump. Photo by Dr. Herbert R. Axelrod at the Berlin Aquarium.

Asprotilapia leptura is the only member of its genus. It is a mouthbrooder and occurs over sand, sifting out the edible organisms. Photo by Pierre Brichard.

Grammatotria lemairei is another sandy bottom mouthbrooder, feeding on the small invertebrates and diatoms it finds there. It grows to over 25 cm in nature. Photo by Pierre Brichard.

Simochromis dardennei grows to about 26 cm but doesn't develop its attractive colors until it is beyond 18 cm. It is also a mouthbrooder. Photo by Dr. Herbert R. Axelrod.

Bathybates ferox is a typical predator on plankton-feeding fishes such as clupeids. It is a large (to 36 cm) mouthbrooder living in the deepest habitable portions of the lake. Photo by Pierre Brichard.

Lamprologus brevis is one of the smallest cichlids in Lake Tanganyika, reaching only about 50 mm, and is a substrate-spawning shell-dweller. Photo by Thierry Brichard.

Lamprologus ornatipinnis is also a substrate–spawning shell-dweller. It is small (to 74mm) and feeds mainly on copepods. Photo by Pierre Brichard.

Lamprologus moorii is currently known to exist in two color phases. This is the yellow morph. Photo by Shuichi Iwai, Midori Shobo.

This is one of the newer species of *Lamprologus* described by Max Poll, *Lamprologus caudopunctatus*. Little is known about this and the above species, but both are expected to be substrate spawners. Photo by Pierre Brichard.

Lamprologus moorii, brown morph. This is an omnivorous species found only in the southern parts of the lake. Photo by Shuichi Iwai, Midori Shobo.

Lamprologus toae inhabits rocky areas where it spawns among the rubble. It is relatively small (up to 100mm) and feeds mainly on aquatic insect larvae. Photo by Pierre Brichard.

Lamprologus callipterus is well established in the hobby already. It occurs over sandy areas and, according to Pierre Brichard, breeds in mollusc shells. Photo by Tetsu Sato, Midori Shobo.

Lamprologus "staecki" comes from the southern coasts of Lake Tanganyika. Photo by Pierre Brichard.

Lamprologus pleuromaculatus does not grow too big (to 11 cm) but is still a predator on small fishes. It is said to breed in a crater-like nest in the sand. Photo by Yuji Suzuki, Midori Shobo.

Lamprologus mondabu is recognizable by its square-cut to almost forked tail. Photo by Dr. Herbert R. Axelrod at the Berlin Aquarium.

Lamprologus elongatus is a predator cruising about over the substrate but not far from the rocks. Small fishes should not be placed in the same aquarium. Photo by H. Scheuermann.

A spawning pair of *Lamprologus elongatus* with their most recent brood. Note the typical rocks-only African lake cichlid tank. Photo by H. Scheuermann.

Lamprologus tetracanthus is a sand-dweller, spawning in a crater-like nest dug in the sand. It guards the fry. Photo by Mitsuyoshi Tatematsu, Midori Shobo.

Lamprologus tetracanthus with a more prominent red stripe in the dorsal fin; the specimen above has the black stripe more prominent. Photo by Pierre Brichard.

Lamprologus furcifer lives in caves in the lake or other dark sheltered areas. It swims in an upside-down position with its belly against the rock. Photo by Glen S. Axelrod.

A spawning pair of *Lamprologus furcifer* in an aquarium. The green eggs are deposited on the side of a rock or in some other sheltered position. Photo by J. Shortreed.

Lamprologus fasciatus has ten spines in the anal fin, more than any other species of *Lamprologus* except *L. compressiceps*. It normally feeds on small invertebrates. Photo by Glen S. Axelrod.

Lamprologus leleupi is one of the few very popular aquarium species of *Lamprologus* from Lake Tanganyika. Most of this popularity is of course due to its brilliant yellow color. Photo by H.-J. Richter.

Lamprologus leleupi are not all bright yellow. This pair may be *L. leleupi melas*, a darker colored subspecies. Photo by Dr. Herbert R. Axelrod.

A golden form of *Lamprologus leleupi*. It has been reported that the diet and/or color of the substrate has much to do with the color of the fish. Photo by Dr. Herbert R. Axelrod.

The current interest in Tanganyika fishes is centered around the shell-dwellers, notably *L. brevis, L. ocellatus, L.* sp. ("magaire" in the trade), *L. ornatipinnis,* and the one pictured here, *L. meeli.*

Lamprologus modestus , a nest-breeder in which both parents guard the fry. It is omnivorous, feeding on various bottom invertebrates. Photo by Dr. Herbert R. Axelrod.

Lamprologus christyi (above and below) feeds on crustaceans, insect larvae and molluscs with the help of a couple of large molar-shaped pharyngeal teeth. It spawns in a crater-like nest. Upper photo by Dr. Herbert R. Axelrod; lower photo by Dr. R.J. Goldstein.

Lamprologus brichardi is perhaps the most popular species of *Lamprologus* as far as aquarists are concerned. The lyretail caudal fin no doubt has something to do with this. Photo by Dr. Herbert R. Axelrod.

Lamprologus savoryi is very closely related to *L. brichardi*. It is nastier in disposition and more secretive by nature. Photo by Pierre Brichard.

A young individual of *Lamprologus brichardi*. The white edges to the fins appear early, and the lyretail is just starting to develop in this fish. Photo by H.-J. Richter.

Lamprologus sp. of the *L. modestus* type. *Lamprologus* is a complicated genus and sorely in need of revision. Photo by Glen S. Axelrod.

Lamprologus mustax is one of the more distinctively marked species of Lake Tanganyika. The fish shown here is about 13 cm in length. Photo by Yasukuki Kurasawa.

Lamprologus niger is a mollusc- and insect-feeder inhabiting sheltered areas among the rocks. So far large numbers of this species have not been found. Photo by Pierre Brichard.

Lamprologus obscurus was also recently described by Poll. More new species of *Lamprologus* are sure to be described in the near future. Photo by Pierre Brichard.

Lamprologus tretocephalus is often confused with *L. sexfasciatus* or even young individuals of *Cyphotilapia frontosa* because of the barred pattern. Photo by G. Meola, African Fish Imports.

Lamprologus sexfasciatus has the most bars, four of which are under the dorsal fin and one immediately before it. Photo by Aaron Norman.

A group of young *Lamprologus tretocephalus* in a basically rock-decorated African cichlid tank. Note only three bars below the dorsal fin. Photo by G. Meola, African Fish Imports.

Lamprologus lemairei is, according to Pierre Brichard, "the most voracious and predatory species in the genus." Photo by Glen S. Axelrod.

Lamprologus prochilus is another predatory species of *Lamprologus* that was described in recent years by Bailey & Stewart. Photo by Pierre Brichard.

Lamprologus compressiceps is one of the more distinctive members of the genus *Lamprologus*. Its greatly compressed body is not evident in these photos taken from the side. Photo by W. Hoppe.

There are several color morphs of *Lamprologus compressiceps* ranging from brown to yellow to red. This is the one commonly seen in the aquarium trade. Photo courtesy of Wardley Products.

A species of *Lamprologus* that is very closely related to *L. compressiceps* is this *L. calvus*. It has fewer anal fin spines, among other things. Photo by Pierre Brichard.

According to Brichard, *Lamprologus compressiceps,* in spite of its large mouth, feeds only on shrimps and crabs—not on small fishes. Photo by Dr. Herbert R. Axelrod.

Lamprologus compressiceps may be found in several color morphs such as the yellow one seen above and the red one seen below. Photos by Glen S. Axelrod.

A spotted form of *Chalinochromis* that may turn out to be a species new to science. Photo by G. Meola, African Fish Imports.

A striped form of *Chalinochromis* has also been found in Lake Tanganyika and it, too, may turn out to be a new species. Photo by G. Meola, African Fish Imports.

Young *Chalinochromis brichardi* start out with solid stripes. These stripes break up into spots and eventually disappear. Photo by Pierre Brichard.

Normal *Chalinochromis brichardi*. Both have the characteristic head bridles, the upper fish still retaining the posterior dark spot in the dorsal fin. Photo by Dr. Herbert R. Axelrod.

Julidochromis dickfeldi is the newest member of the genus and belongs to the so-called "dwarf" group comprised of *J. ornatus* and *J. transcriptus*. Photo by Pierre Brichard.

Julidochromis dickfeldi, like other members of the genus, spawns generally in protected rocky areas such as caves, crevices, etc. Photo by Dr. Herbert R. Axelrod.

Julidochromis ornatus was one of the early importations and has become an established favorite among rift lake cichlid enthusiasts. Photo by H. Hansen.

Julidochromis transcriptus is an attractive species, especially the melanistic forms such as this one. Photo by Thierry Brichard.

The southern form of *Julidochromis regani*. *J. regani* and *J. marlieri* are the larger species of the genus, reaching a length of 15 cm compared with 6 to 7 for the dwarf forms. Photo by Glen S. Axelrod.

A more typical *Julidochromis regani*. The horizontal stripes and barred caudal fin make it easily recognizable. Photo by Thierry Brichard.

Probably the second most popular species of *Julidochromis* is *J. marlieri;* at least it is readily available in most stores that sell African lake cichlids. Photo by R. J. Goldstein.

Julidochromis marlieri is an excellent aquarium fish that can be readily spawned. Finding a true pair may be somewhat difficult, however. Photo courtesy Wardley Products Co.

This specimen of *Julidochromis* appears intermediate between *J. ornatus* and *J. transcriptus,* casting some doubt as to the validity of the latter species. Photo by H. Hansen, Aquarium Berlin.

There are other varieties of *Julidochromis* which are hard to place with any particular species. This one may be a pale form of *J. marlieri.* Photo by Thierry Brichard.

The dark band in the dorsal fin is a characteristic of *Simochromis marginatus*. Species in this genus are closely related to the genus *Tropheus*. Photo by Glen S. Axelrod.

Simochromis curvifrons has been placed in its own genus (*Pseudosimochromis*) by Nelissen, but this move has not been generally accepted. Photo by Pierre Brichard.

Simochromis diagramma is one of the larger members of the genus (to 20 cm) and is found throughout the lake where there are rocks and rubble. Photo by Dr. Herbert R. Axelrod.

Simochromis babaulti is much smaller (to 10 cm) and often found in water as shallow as 20 to 30 cm. Photo by Dr. Herbert R. Axelrod.

A variety of *Simochromis babaulti* with an unusual pattern. Photo by Pierre Brichard.

A colorful variety of a recently described species of *Simochromis* called *S. pleurospilus.* It is a small species (to 6 cm) living in rubble areas in shallow water. Photo by Pierre Brichard.

Tropheus duboisi juveniles are very attractive with white spots on a black background. Photo by Peter Chlupaty.

Tropheus duboisi adult with the characteristic white band about mid-body and no white spots remaining. Photo by Glen S. Axelrod.

Two varieties of *Tropheus duboisi*, one called the narrow olive band (above) and the other called the wide olive band (below). Photos by Glen S. Axelrod.

A variety of *Tropheus moorii* called the yellow-belly. Photo by Glen S. Axelrod.

A dark variety of *Tropheus duboisi* found around Kigoma in Tanzania. Photo by Pierre Brichard.

The "blue-eyed" *Tropheus* from the Nyanza-Lac area of Lake Tanganyika has been described as *Tropheus brichardi* in honor of Pierre Brichard's efforts in the lake. Photo by Pierre Brichard.

Another, and perhaps more common, variety of *Tropheus brichardi* is this striped form. Photo by Dr. Herbert R. Axelrod.

Probably one of the best known fishes from Lake Tanganyika is *Tropheus moorii*. It is now well established as an aquarium favorite. Photo by Dr. Herbert R. Axelrod.

A juvenile red-black *Tropheus moorii* from the northern end of the lake. This is one of the most popular varieties with aquarists. Photo by Pierre Brichard.

A similarly patterned variety of *Tropheus moorii* but with more orange on the body and hence called the orange-black variety. Photo by Pierre Brichard.

An adult red-black variety of *Tropheus moorii*. This species has a number of color varieties throughout the lake. Photo by Pierre Brichard.

One of the red varieties of *Tropheus moorii* (above) and a yellow
form (below). The odd-looking snout seems to be real and not the
result of some damage. Photos by Glen S. Axelrod.

Some of the more colorful varieties of *Tropheus moorii*. The one above is called the orange variety and the one below the red variety or the Chipimbi Cape variety. The red band on the caudal peduncle is real. Photos by Pierre Brichard.

The lunate caudal fin makes the new *Tropheus polli* more easily recognizable. Photo by Glen S. Axelrod.

Petrochromis fasciolatus as well as other members of this genus have many rows of tricuspid teeth for scraping algae. They have been compared to *Petrotilapia tridentiger* of Lake Malawi, which has similar dentition. Photo by Glen S. Axelrod.

Petrochromis polyodon is probably the best known of the *Petrochromis* to the hobbyist. The above fish with shorter fins is probably a female, the one below is a male. Photos by Dr. Herbert R. Axelrod.

A young *Petrochromis orthognathus* from the southern end of the lake. Photo by Pierre Brichard.

An adult *Petrochromis orthognathus*. Note the unusual placement of the anterior orange spots ("egg spots") on the anal fin spines. Photo by Pierre Brichard.

The species of *Petrochromis* are, at this time, very difficult to distinguish and more research (including possible descriptions of new species) is needed to straighten them all out. Until then there will be several specimens that must be called simply *Petrochromis* sp. as these two are. Photos by Glen S. Axelrod.

Petrochromis sp., young adult above and adult below. The species of *Petrochromis* are always found over rocky areas where they scrape algae from the surface of the rocks. They are mouthbrooders producing up to 50 eggs per spawning. Photos by Glen S. Axelrod.

Petrochromis sp. This individual may represent a new species of *Petrochromis*. Photo by Glen S. Axelrod.

A young *Boulengerochromis microlepis*. This species is one of the largest cichlids in the world, attaining lengths of up to 90 cm. It is a substrate spawner digging large nests in the sand. Photo by Pierre Brichard.

Xenotilapia melanogenys develops very beautiful spawning colors. Unfortunately it is delicate and difficult to ship and will not likely appear for sale very often. Photo by Aaron Norman.

Reganochromis calliurum lives on sandy bottoms in the deeper waters of Lake Tanganyika. This beautiful fish is rarely seen in the hobby. Photo by Pierre Brichard.

Haplochromis similis   B. Kahl

Haplochromis chrysonotus   R. Stawikowski
Haplochromis labrosus   B. Kahl

Haplochromis chrysonotus   B. Kahl

Haplochromis ahli   H. Linke
Haplochromis ahli   H. Linke

Haplochromis chrysonotus   R. Stawikowski

Haplochromis ahli   B. Kahl
Haplochromis ahli   R. Stawikowski

Gephyrochromis moorii   R. Stawikowski

Haplochromis moorii   B. Kahl
Haplochromis electra   R. Stawikowski

Haplochromis moorii   B. Kahl

Haplochromis moorii   B. Kahl
Haplochromis sp. cf. *incola*   B. Kahl

Aulonocara nyassae var.   H. Mayland

Aulonocara nyassae var.   H. Mayland
Aulonocara nyassae var.   D. Schaller

Aulonocara nyassae var.   R. Stawikowski

Aulonocara baenschi   H. Mayland
Aulonocara stuartgranti   H. Linke

Haplochromis livingstonii   B. Kahl

Haplochromis ornatus   R. Stawikowski
Aulonocara nyassae   R. Stawikowski

Pseudotropheus sp.cf. *heteropictus*   R. Stawikowski

Labeotropheus fuelleborni   R. Stawikowski
Pseudotropheus brevis   B. Kahl

Pseudotropheus sp. cf. *elongatus*   B. Kahl

Pseudotropheus sp. "teyoi"   R. Stawikowski
Labeotropheus fuelleborni   B. Kahl

Pseudotropheus microstoma   B. Kahl

Pseudotropheus sp. aff. *zebra*   H. Linke
Pseudotropheus crabro   R. Stawikowski

Pseudotropheus zebra   B. Kahl

Pseudotropheus zebra   B. Kahl
Pseudotropheus lombardoi   H. Linke

Pseudotropheus zebra   R. Stawikowski

Pseudotropheus greshakei   H. Mayland
Pseudotropheus zebra   B. Kahl

Pseudotropheus microstoma   B. Kahl

Pseudotropheus macrophthalmus   B. Kahl
Pseudotropheus "eduardi"   B. Kahl

Cyathopharynx furcifer   Schuphe

Ophthalmotilapia ventralis ventralis   Schuphe
Ophthalmotilapia nasutus   Schuphe

Lamprologus kendalli   D. Schaller

Lamprologus buescheri   H. Mayland
Lamprologus fasciatus   H. Mayland

Boulengerochromis microlepis   H. Mayland

Telmatochromis caninus   B. Kahl
Lobochilotes labiatus   B. Kahl

Lamprologus meeli   H. Linke

Lamprologus sp. "magarae"   H. Linke
Lamprologus compressiceps   D. Schaller

Lamprologus multifasciatus   H. Mayland

Lamprologus sp. "magarae"   H. Mayland
Lamprologus ornatipinnis   H. Mayland

Lamprologus brichardi   H. Mayland

Lamprologus brichardi   D. Schaller
Lamprologus cunningtoni   R. Stawikowski

Lamprologus brichardi (daffodil)   H. Mayland

Lamprologus brichardi (daffodil)   D. Schaller
Lamprologus savoryi   H. Mayland

Lamprologus sp. aff. *elongatus*   H. Mayland

Cunningtonia longiventralis   D. Schaller
Callochromis macrops   H. Mayland

Spathodus marlieri   H. Mayland

Haplotaxodon microlepis   H. Mayland
Xenotilapia ochrogenys   H. Mayland

Lamprologus sexfasciatus   D. Schaller

Lamprologus sp. "cylindricus"   H. Mayland
Lamprologus toae   H. Mayland

Lamprologus sexfasciatus (gold)   D. Schaller

Lamprologus tretocephalus   R. Stawikowski
Lamprologus sp. "cylindricus"   D. Schaller

Lamprologus mustax   H. Mayland

Lamprologus attenuatus   R. Stawikowski
Lamprologus leleupi longior   H. Linke

Tropheus polli   R. Stawikowski

Tropheus moorii var.   R. Stawikowski
Simochromis diagramma   B. Kahl

Tropheus moorii kasabae   R. Stawikowski

Tropheus moorii   D. Schaller
Petrochromis fasciolatus   R. Stawikowski

Julidochromis regani   B. Kahl

Julidochromis ornatus   H. Linke
Julidochromis ornatus   B. Kahl

Julidochromis marlieri   B. Kahl

Julidochromis regani   B. Kahl
Julidochromis regani   H. Linke

Julidochromis dickfeldi   B. Kahl

Chalinochromis brichardi   H. Mayland
Cyprichromis leptosoma   D. Schaller

Julidochromis marlieri   H. Linke

Cyrichromis brieni   B. Kahl
Cyrichromis microlepidotus   D. Schaller

Xenotilapia spilopterus from the southern end of the lake. The large black spot in the middle of the dorsal fin helps identify it. Photo by Pierre Brichard.

Xenotilapia longispinis longispinis. Members of this genus are mouthbrooders that feed on bottom invertebrates by sifting through the sand. Photo by Glen S. Axelrod.

This is a new species of *Xenotilapia* from the northern part of
Lake Tanganyika. Photo by Pierre Brichard.

Xenotilapia ochrogenys ochrogenys is very sensitive to changes
in water quality, etc. and therefore extremely difficult to ship.
Photo by Pierre Brichard.

Xenotilapia sima grows to more than 15 cm. It is not as colorful as many other Tanganyikan cichlids and thus rarely imported. Photo by Dr. Herbert R. Axelrod.

The colors of *Xenotilapia boulengeri,* like many other sand-dwellers, blend well with their backgrounds making them difficult to see. They also suffer from the bends, more reason for them being avoided by collectors and shippers. Photo by Dr. Herbert R. Axelrod.

Callochromis macrops melanostigma is generally found on sandy bottoms, often near the surf zone. It is more colorful than the other subspecies, *C. macrops macrops*. Photo by Dr. Herbert R. Axelrod.

Callochromis pleurospilus, the second of two species in the genus, is quite attractive with its myriad of pastel colors. Like *C. macrops*, it is a sand-sifter. Photo by Dr. Herbert R. Axelrod.

Callochromis pleurospilus, female above and male below. Eggs are reported by Pierre Brichard as "salmon-pink, pear-shaped and about 2-3 mm long." Photos by Dr. Herbert R. Axelrod.

The teeth of this *Perissodus paradoxus* are modified to enable it to dislodge scales from its victims. This scale-eater is not recommended for community aquariums. Photo by Dr. Herbert R. Axelrod.

Perissodus microlepis is another scale-eater from Lake Tanganyika. It inhabits rocky slopes waiting for victims such as *Tropheus moorii* to stray within range. Photo by Pierre Brichard.

This *Perissodus straeleni* has an unusual pattern. It is not unique, as other individuals with the same black head mask were seen. Photo by Glen S. Axelrod.

The more usual color pattern of *Perissodus straeleni*. It mimics other barred species such as *Cyphotilapia frontosa,* perhaps to enable it to get close enough to snatch some scales. Photo by Pierre Brichard.

Lobochilotes labiatus, head study above, side view below. This individual is only partially grown. *L. labiatus* is one of the larger cichlids in the lake, growing to 40 cm. Photos by Dr. Herbert R. Axelrod.

Lobochilotes labiatus feeds on invertebrates such as molluscs and crabs, the enlarged lips possibly aiding them in finding this prey. Photo by G. Marcuse.

An adult *Lobochilotes labiatus* with the lips fully developed. This species is a mouthbrooder that comes up to shallow water to brood and release the fry. Photo by Glen S. Axelrod.

Sarotherodon tanganicae (above and below) has been able to spread out into the lake rather than being restricted to estuaries, etc., like most *Sarotherodon* or *Tilapia* species. It feeds on the rocky biocover. Photos by Glen S. Axelrod.

Haplochromis benthicola is a deep-living rock-dweller. The male (shown here) is rather somber-colored, but the female is said to be red-gold and black. Photo by Pierre Brichard.

Tilapia rendalli is one of the relatively few non-endemic cichlids of Lake Tanganyika. Note the *"Tilapia* spot" in the dorsal fin. Photo by Dr. Herbert R. Axelrod.

Haplochromis pfefferi was formerly placed in the genus *Limnochromis*. It is a mouthbrooder that may be found, though not in great numbers, throughout the lake. It is usually solitary, pairing only for reproduction. Photo above by Dr. Herbert R. Axelrod; photo below by Pierre Brichard.

Haplochromis horei (male above, female below) is quite distinctly (and attractively) patterned. It is found in shallow water (less
than one meter) and is a predator on fishes. Photos by Glen S.
Axelrod.

Haplochromis burtoni (above and below) inhabits coastal delta swamps rather than the open waters of Lake Tanganyika. It has a string of egg spots in the anal fin which are shown in the close-up below. Photos by Glen S. Axelrod.

Tylochromis lateralis, young specimen above, adult below.
Species of this genus are thought to be maternal mouthbrooders;
that is, the female cares for the eggs. Photo above by Dr. Herbert
R. Axelrod; photo below by H. Hansen, Aquarium Berlin.

Telmatochromis caninus grows a little larger than *T. temporalis*, 12 cm to 10 cm respectively. Both these species seem quite distinct from *T. bifrenatus* and *T. vittatus*, these latter species possibly being placed in their own genus at some future date. Photo by Dr. Herbert R. Axelrod.

Telmatochromis caninus spawning. The male is in front. Note the clutch of eggs in the rock cave in the background. Photo by H.-J. Richter.

Telmatochromis bifrenatus inhabits rocky areas and is able to find shelter very quickly when danger threatens. Photo by H.-J. Richter.

Telmatochromis bifrenatus has conspicuous diagonal lines crossing the longitudinal band. Photo by Dr. Herbert R. Axelrod.

Limnochromis auritus is a common species in Lake Tanganyika but a rather difficult one to ship. It does well in the aquarium and can be purchased from time to time. Photo by Dr. Herbert R. Axelrod at the Berlin Aquarium.

Limnochromis auritus reaches a length of about 14 cm and apparently is a mouthbrooder. Photo by Dr. R. J. Goldstein.

One of the goby cichlids, *Spathodus marlieri,* is a better swimmer than the rest and can be seen swimming over the rocks for some distance. Photo by Glen S. Axelrod.

Spathodus erythrodon lives in shallow water, hopping from rock to rock like marine gobies, hence the name goby cichlid. Photo by Dr. Herbert R. Axelrod.

Astatoreochromis straeleni is another swamp-dweller and often lives alongside *Haplochromis burtoni* in the muddy lagoons and estuaries. Photo by Pierre Brichard.

Astatoreochromis straeleni was once considered a *Haplochromis* but was moved by Max Poll to *Astatoreochromis*. Photo by Aaron Norman.

Eretmodus cyanostictus is another of the goby cichlids. It also lives in shallow water among the rocks from which they feed. Photo by Glen S. Axelrod.

Eretmodus cyanostictus is also a mouthbrooder, incubating about 25 pink eggs per spawning. Photo courtesy Wardley Products Co.

A variety of *Spathodus erythrodon* from the southern end of the lake. This species seldom exceeds 50 mm total length. Photo by Pierre Brichard.

Spathodus erythrodon is a mouthbrooder and feeds on microorganisms in the rock biocover. Photo by Dr. Herbert R. Axelrod.

Tanganicodus irsacae (above and below), the final member of the goby cichlids. The different genera of goby cichlids are most easily distinguished by tooth type. In *Tanganicodus* they look most like those of the Malawian genus *Labidochromis*. Photos by Glen S. Axelrod.

Triglachromis otostigma is easily recognized by the narrow slanting silvery stripes on the sides. Photo of a young male by Dr. Herbert R. Axelrod.

The lower rays of the pectoral fins of *Triglachromis otostigma* are independently movable and may be used in the detection of food. Photo by Homer Arment.

Telmatochromis vittatus is very similar in appearance to the previous species but has a more sloping snout. Photo by Pierre Brichard.

Telmatochromis bifrenatus, typical form with the additional dark stripe between the lateral and dorsal base stripes. Photo by Glen S. Axelrod.

Telmatochromis temporalis was among the first fishes exported from Lake Tanganyika. It is still available although its popularity has decreased somewhat. Photo by Dr. Herbert R. Axelrod.

Telmatochromis temporalis is usually distinguishable from *T. caninus* by the blue streak below the eye. Photo by Aaron Norman.

ILLUSTRATIONS INDEX